SOUTHERN MAUNSELL

4-4-0 CLASSES (L, D1, E1, L1 AND V)

SOUTHERN MAUNSELL

4-4-0 CLASSES (L, D1, E1, L1 AND V)

DAVID MAIDMENT

First published in Great Britain in 2019 by
PEN & SWORD TRANSPORT
An imprint of
Pen & Sword Books Ltd
Yorkshire - Philadelphia

ISBN 978 1 52671 469 5

A CIP catalogue record for this book is available from the British Library.

Typeset in Palatino by Aura Technology and Software Services, India.
Printed and bound in India by Replika Press Pvt Ltd.

Pen & Sword Books Ltd incorporates the Imprints of Pen & Sword Books Archaeology, Atlas, Aviation, Battleground, Discovery, Family History, History, Maritime, Military, Naval, Politics, Railways, Select, Transport, True Crime, Fiction, Frontline Books, Leo Cooper, Praetorian Press, Seaforth Publishing, Wharncliffe and White Owl.

For a complete list of Pen & Sword titles please contact

PEN & SWORD BOOKS LIMITED
47 Church Street, Barnsley, South Yorkshire, S70 2AS, England
E-mail: enquiries@pen-and-sword.co.uk
Website: www.pen-and-sword.co.uk

Or
PEN AND SWORD BOOKS
1950 Lawrence Rd, Havertown, PA 19083, USA
E-mail: Uspen-and-sword@casematepublishers.com
Website: www.penandswordbooks.com

All David Maidment's royalties from this book will be donated to the Railway Children charity [reg. no. 1058991] [www.railwaychildren.org.uk]

Other books by David Maidment:

Novels (Religious historical fiction)
The Child Madonna, Melrose Books, 2009
The Missing Madonna, PublishNation, 2012
The Madonna and her Sons, PublishNation, 2015

Novels (Railway fiction)
Lives on the Line, Max Books, 2013

Non-fiction (Railways)
The Toss of a Coin, PublishNation, 2014
A Privileged Journey, Pen & Sword, 2015
An Indian Summer of Steam, Pen & Sword, 2015
Great Western Eight-Coupled Heavy Freight Locomotives, Pen & Sword, 2015
Great Western Moguls and Prairies, Pen & Sword, 2016
Southern Urie and Maunsell 2-cylinder 4-6-0s, Pen & Sword, 2016
Great Western Small-Wheeled Double-Framed 4-4-0s, Pen & Sword, 2017
The Development of the German Pacific Locomotive, Pen & Sword, 2017
Great Western Large-Wheeled Double-Framed 4-4-0s, Pen & Sword, 2017
Great Western Counties, 4-4-0s, 4-4-2Ts & 4-6-0s, Pen & Sword, 2018
Southern Railway, Maunsell Moguls and Tank Locomotive Classes, Pen & Sword, 2018

Non-fiction (Street Children)
The Other Railway Children, PublishNation, 2012
Nobody ever listened to me, PublishNation, 2012

Cover photo:
31507 at London Bridge on the 12.44pm parcels train to Ramsgate, c1960 R.C. Riley Collection

Back cover:
31545 at Stewarts Lane, with a Maunsell 'W' 2-6-4T behind, 10 May 1959 R.C. Riley

30901 'Winchester' ex works in BR lined green livery at Ashford, 20.6.1960. R.C.Riley

30914 *Eastbourne* in BR lined black mixed traffic livery at Stewarts Lane, 24 May 1958 R.C. Riley

CONTENTS

PREFACE & ACKNOWLEDGEMENTS

After writing three books about 4-4-0s on the Great Western Railway, Pen & Sword suggested I turn my attention to the Southern Railway, or at least its constituent parts, in particular the old much maligned South Eastern & Chatham Railway. Not, however, to the contemporaries of the GWR engines I described, the SE&CR 'Ds' and 'Es', but to the effective rebuilding of them by Richard Maunsell as 'D1s' and 'E1s'. I then considered including the Wainwright 'Ls', the design of which was modified and completed by Maunsell, and later, the so-called 'improved Ls', the 'L1s' produced by Maunsell urgently to tackle problems in the early days of the regrouped Southern Railway to bridge the gap until track improvements were made and larger locomotives could be developed and utilised. Finally, I agreed to tackle the whole realm of Maunsell's 4-4-0 classes by including his masterpiece, the 'Schools'. Much has already been written about these engines, but I have been able to draw together through my research their operation and performance as well as technical descriptions of the class from writers such as Cecil J. Allen, O.S. Nock, Ronald Nelson, S.C. Townroe and D.W. Winkworth, and add my own personal recollections of them in the late 1950s when I was a regular traveller between Woking and Waterloo with the occasional sortie to Victoria or Charing Cross before electrification there in 1959 and the early 1960s.

I am once again indebted to the photographers or owners of collections of photographs who have permitted me to publish many of the best images from their collections without charge or at a reduced fee as all the royalties from the book will be donated, as with all my books, to the Railway Children charity that I founded in 1995 to support and rehabilitate street children found around railway and bus stations in India and East Africa and runaway children in the UK. My grateful thanks therefore in particular go to the Manchester Locomotive Society (MLS) and Paul Shackcloth for the use of photographs from their comprehensive archive; Mike Bentley, a member of the Society for access to, and use of, the many photographs in his collection; John Scott-Morgan who found some additional photos for me; and Rodney Lissenden who holds the colour slide collections from the late Dick Riley, David Clark and Ken Wightman on behalf of their families. I am able to add a few of my own photographs, particularly of the 'Schools' class, when I was familiar with them between 1957 and 1962.

I had my own personal association with one of them as after initial education at Surbiton County Grammar School, where, after lessons one day in 1950, I 'copped' my first 'Schools' (30917 *Ardingly*), I was fortunate to obtain a scholarship through the Surrey County Council 'assisted places' scheme to the famous public school, Charterhouse. I thus had a special relationship with 30903, a model of which stands beside me on the windowsill of my study as I type these words. I realise now the debt I owed that County Council, for not only did they fund my schooling there, but paid for my annual season ticket from Woking to Waterloo (£33 in 1958!) as part of my grant to support me whilst at University College London, and thus my three years of commuting behind steam – with hundreds of journeys behind Maunsell's 'Schools', 'King Arthurs' and 'Lord Nelsons' as well as Bulleid pacifics and BR Standard 4-6-0s.

Finally, I would like to acknowledge the support and professional help I've received from Pen and Sword – John Scott-Morgan, Commissioning Editor; Janet Brookes, Production Manager of the Transport imprint; Carol Trow, my regular editor; and the design team that the company has used for my books. They do a very professional job for which I'm grateful and helped me to maximise the amount I can give to the charity which is now described by the United Nations as the largest charity in the world working exclusively for street children (in the context I think of a charity *only* raising $5million a year being the largest for the millions of street children worldwide who are so vulnerable and need much more help). We have plans to increase that support to £8 million a year over the next five years with a substantial proportion being raised by both the professional and heritage railway industries, railway employees and railway enthusiasts whom I wish to acknowledge and thank.

David Maidment
www.davidmaidment.com
www.railwaychildren.org.uk
March 2019

INTRODUCTION

The South Eastern Railway and the London, Chatham & Dover Railway finally buried the hatchet by beginning to work together instead of engaging in competition that threatened to bankrupt both of them at the beginning of 1899, and it was formalised on 5 August of that year. Both companies retained their Board of management, but pooled resources, and the initial South Eastern & Chatham Companies' Managing Committee soon became the South Eastern & Chatham Railway (SE&CR). Victoria station was rebuilt and the docks at Dover and Folkestone were enlarged.

Until 1898, the South Eastern routes had a speed limit of 60mph, although the more difficult LC&DR route allowed higher speeds – achieved mainly on downhill sections in the Swanley Junction–Sole Street area. Things relaxed a bit after the turn of the century and Rous-Marten recorded 75mph in the Hildenborough area in a 1904 *Railway Magazine* article. There was one seven hour link from London to Paris using the 2.20pm Victoria–Paris via Folkestone and Boulogne, but the majority of trains up to the First World War averaged no more than 54mph on the most favourable stretches. The fastest boat trains could only manage 45mph average London to Dover and on the Hastings line 40mph was the maximum average. All rolling stock was composed of non-corridor vehicles (and remained so up to the Grouping in 1923 apart from continental boat trains from 1921).

Harry Wainwright, previously the South Eastern Railway's Carriage & Wagon Superintendent, was appointed Locomotive, Carriage & Wagon Superintendent of the combined company at the end of 1898, and his lack of steam locomotive design experience was balanced by the decision to appoint the LC&DR's chief draughtsman, Robert Surtees, to the same position in the amalgamated railway company. Initially, locomotives of the new company were built and repaired at both Longhedge and Ashford Works, although in 1906 all new construction took place at Ashford and the Erecting and Boiler Shops were extended. Longhedge Works closed at the end of 1911 and its relatively modern equipment was rehoused at Ashford. At the start of the amalgamated company, it inherited 459 locomotives from the South Eastern Railway and 215 from the Chatham company. The SER engines retained their numbers and the LC&DR engines had 459 added to their numbers.

Surtees was a competent steam engineer and had designed simple straightforward reliable locomotives and he continued as the backbone to the new regime, with Wainwright maintaining his carriage and wagon expertise and applying the finishing touches to Surtees' designs – especially their aesthetics and livery. There was a need to replace ageing designs of both companies and initially five 4-4-0s were purchased from the Great North of Scotland Railway, while Surtees and Wainwright busied themselves, trying to increase power but hampered by civil engineering weight restrictions, especially on the LC&DR routes. Wainwright conceived the need for a moderately powerful 4-4-0 for general main line work and an outside cylinder atlantic for heavier boat expresses. However, Surtees convinced him that a 4-4-0 of similar type to the Caledonian's successful 'Dunalastair' series would be cheaper and enable more standardisation – a definite need in the motive power department at that time. The argument was won when the Civil Engineer objected to the atlantic design between Tonbridge and Hastings and more importantly, over the whole Chatham route to Dover.

A 4-4-0 was therefore adapted to cater for both needs and twenty engines were ordered in April 1899, ten to be built at Ashford and ten by a private manufacturer. A tender from Sharp, Stewart at £3,575 each was accepted for delivery by the end of 1900. Subsequently, a further twenty were ordered, five from Vulcan Foundry at £3,345, five from Robert Stephenson & Co at £3,320 and ten from Dübs & Co at £3,370 – the cost of all significantly higher than the £2,950 finally authorised for the in-house Ashford build. This class was the well-known stylish class 'D', of which one example, 737, built in 1901 at Ashford Works, was preserved as part of the collection at the National Railway Museum at York. These engines had two 19in x 26in inside cylinders, 6ft 8in coupled wheels, heating area of 1,505sqft, 175lb boiler pressure, axle-loading of 17 tons, and a total weight with tender of 88 tons 5 cwt. The tender's capacity was for 3,450 gallons of water and 4 tons of coal.

They were immediately employed on both parts of the system, with the South Eastern Section based engines at Bricklayers Arms, Cannon Street, Ashford, Ramsgate and Dover Town, and the locomotives for the Chatham Section allocated to Battersea (later Stewarts Lane), Dover Priory and Margate. The locomotives were all delivered between 1901 and 1903 (the Vulcan and Stephenson engines missed their promised delivery dates). The boiler proved free-steaming and the riding was excellent. They climbed well and there were few hot boxes. There was – after a few months' service – some concern over leaking tubes and priming, but the main cause was found to be the hard Kentish water and depot water softening plants solved the problem.

A further batch of eleven was built at Ashford in 1906/7, but the introduction of steam-heating on the most important expresses increased the demand on the engine and the Surtees/Wainwright team designed a 4-4-0 with 6ft 6in coupled wheels and 180lb boiler pressure and Belpaire firebox, the 'E' class, twenty of which were constructed between 1906 and 1909 and were diagrammed to the heaviest expresses. Hard-working engines with poor quality coal tended to throw the fire and there were many complaints from farmers with fields bordering the railway. One of the 'Ds', 247, was modified with an extended smokebox (a strange design, the diameter being the same size as

Ashford built 488, 'D' class 4-4-0, delivered in 1902 and allocated to Dover Town, seen here at Cannon Street, 1902. F. Moore/MLS Collection

'D' 247 fitted experimentally in 1907 with a small diameter extension to the smokebox, although by 1914 it had a conventional extended smokebox, which it retained until rebuilt as a 'D1' in 1921. It was the only 'D' fitted with an extended smokebox, c1907. J.M. Bentley Collection

'E' 273, the prototype in Works grey, built at Ashford in 1906 before being fitted with an extended smokebox.
Official SR photo/J.M. Bentley Collection

'E' 175 with extended smokebox at Bricklayers Arms, c1923. 175 was not fitted with an extended smokebox until after the First World War.
J.M. Bentley Collection

the smokebox door and not the boiler). It was clearly successful, as all the new 'Es' were so equipped although with a more conventional shape matching the boiler diameter. 247 retained its unique look until rebuilt as a 'D1'.

Loads were increasing and Wainwright was still attempting to persuade the Civil Engineer to allow heavier locomotives and produced a number of designs including 4-6-0s between 1907 and 1913, but his entreaties were unsuccessful. Surtees and Wainwright then produced a 4-4-0 with greater power, the 'L' (see chapter 3). However, before these could be built, the SE&CR underwent a critical period in 1912/3 when there were severe locomotive problems and failures, some of which occurred because of the directors' insistence of the closure of Longhedge Works before the extensions at Ashford were completed and it could fully take over the work of both Shops. Harry Wainwright had managed the initial merger of the two companies that formed the SE&CR well but having responsibility for the design and construction of locomotives, carriages and wagons and also the running in traffic of the SE&CR locomotive fleet and rolling stock was really too much for one man. He seems to have had some heart problems, possibly caused by the stress of work, suffered a painful divorce when his wife left him for a millionaire and things started going wrong, particularly at Ashford Works, perhaps the result of the directors' decision. Wainwright seems to have borne the brunt of the criticism rather than the directors and he was forced to retire in 1913 at the age of only 48, and died prematurely only twelve years later. Maunsell was appointed as described in the following chapters.

Chapter 1

MAUNSELL IN IRELAND

Richard Maunsell was born at Raheny, County Dublin, in Ireland on 26 May 1868. He came from an old family dating back to Norman times which had settled in Ireland in the 1340s. His predecessors were land-owners and were in the legal profession, but from an early age the boy showed his primary interest to be engineering. He was one of a large family – six brothers and four sisters – and attended a Public School, the Royal School at Armagh, in 1882, before training after pressure from his father for a Law degree at Trinity College, Dublin, in 1886. However, the backbone BA course at the university was followed by all students and he was able to specialise in engineering and was also to benefit from his solicitor father's contacts with the Board of the Irish Great Southern & Western Railway, becoming simultaneously a pupil of H.A. Ivatt at its Inchicore Works.

Maunsell managed to combine the continuation of his Trinity college studies with his Inchicore apprenticeship, as well as becoming an outstanding sportsman in both athletics and cricket, which he played for his college. In 1891, he completed his BA course at Trinity and left Inchicore for a year's experience under John Aspinall of the Lancashire & Yorkshire Railway, the result of a close Ivatt-Aspinall relationship. Clearly the engineering establishment already had the view that Richard Maunsell was a potential 'high-flyer'. Maunsell worked at the newly opened Horwich Works where Aspinall's successful 2-4-2Ts were currently under construction. Alongside Maunsell were three other young men who later became CMEs, Henry Fowler, Henry Hoy and George Hughes, and Maunsell retained contact with these later, especially Henry Fowler, after they'd both become CMEs of two of the 'Big Four' in 1923.

Maunsell had a number of basic shed appointments in the Blackpool and Fleetwood District after experience on the design side, and during this time, was courting Edith Pearson, who he had met during his contacts with the Aspinall family. However, Edith's father prevented their engagement until Maunsell was better able to assure him of his career earning prospects, so Maunsell sought a higher paid post and successfully applied to be Assistant District Locomotive Superintendent of the East India Railway based at Jamalpur at a salary of Rs480 a month (approximately £323 a year) which was nearly double his salary on the L&YR. The East India Railway was very extensive, the second largest railway in India. Maunsell gained rapid promotion there, and after a spell at Tundla on the Allahabad–Delhi route, was transferred back to Jamalpur, then as the Principal District Locomotive Superintendent.

However, even the increased salaries available to Maunsell were insufficient to persuade Edith's father to endorse the proposed marriage. It was only on Patrick Stirling's death, when H.A. Ivatt was appointed to the senior engineering post of the Great Northern Railway, leaving the Locomotive Superintendent's post of Ireland's GS&WR for his former Assistant, Robert Coey, and the latter, his vacated position and free house open for Maunsell. He was therefore appointed at the young age of twenty-eight to the post of Assistant Locomotive Engineer and Works Manager in March 1896, and Richard and Edith were married in June of that year.

Maunsell immediately set about the reorganisation and modernisation of Inchicore Works and between 1897 and 1902, the GS&WR increased its network size by 70 per cent through company take-overs, with Inchicore becoming responsible for the replacement and maintenance of the increased locomotive fleet. Maunsell was a close observer and protagonist of Coey's locomotive developments in the first decade of the new century, including 4-4-0s and 2-6-0s, and experiments with superheating, until Coey was forced to resign in 1911 because of increasing ill health. Richard Maunsell was appointed Locomotive Superintendent on 30 June 1911. Edward Watson was appointed to Maunsell's previous post, after having been Assistant Works Manager at Swindon from where he brought knowledge of the developments being pioneered by Churchward. Maunsell designed just one tender 4-4-0 at Inchicore in 1912, a development of a Coey initiated design, but with a Belpaire firebox, Schmidt superheater and a higher boiler pressure and an 0-6-0 goods engine (the '257' class), the building of which had been commenced before he was appointed and moved to Ashford as the Locomotive, Carriage & Wagon Superintendent of the South Eastern & Chatham Railway.

The 4-4-0 was named *Sir William Goulding* after the current Chairman of the Company. Although much of the groundwork of the design was initiated by Coey, the use of the Belpaire firebox, the Walschaerts valve gear and the improved cab design seem to have been inspired by Maunsell. It was also the first engine of the company to be designed with a superheater, although trials had been undertaken with an existing Coey 4-4-0, No.326. The two inside cylinders were 20in by 26in stroke, the total heating surface was 1,855sqft (including superheater heating surface of 335sqft) and the grate area was 24.8sqft. The boiler was designed for 160lbs psi pressure although in later days this was increased to 175lbs. Further new refinements included a mechanical lubricator, steam sanding and raised running plate over the coupled wheels making the drivers' preparation tasks easier. No.341 was the most powerful passenger engine in Ireland when built but had a high axle-load of 19 tons 2 cwt which restricted its use to the Kingsbridge and Cork main line. Maunsell had obviously intended this to be the first of a series but his successor did not proceed with the design. Although well-liked by crews, as a solitary example it was non-standard and was withdrawn in 1928.

Chapter 2

MAUNSELL AT ASHFORD AND EASTLEIGH

Maunsell's reign at Inchicore was short-lived, for in 1913 he was approached by the South Eastern & Chatham railway seeking a replacement for their Locomotive, Carriage and Wagon Engineer, Harry Wainwright, who took early retirement. He was appointed in December 1913 and had the immediate task of reorganising Ashford Works which could not cope with the workload then placed on it, and the organisation of which had become a mess – Wainwright had been a fine engineer but his management skills, especially in later years, were weak. Maunsell brought his key skills of administration and efficient management, and his engineering knowledge required to tackle the need for greater power to manage the increasing passenger loadings on the SE&CR. The immediate need was met by the already ordered 'L' class 4-4-0s, ten of which were supplied by Borsig's of Berlin and twelve direct from Beyer, Peacock's Works in Manchester to alleviate the shortcomings at Ashford.

Many of Wainwright's team, including Surtees, were nearing retirement age and Maunsell soon assembled a new and very competent team, including James Clayton from Derby and George Pearson and Harry Holcroft from Swindon, although the onset of the First World War restricted their immediate influence. Maunsell was strong enough to bring about significant changes – partly to counter the influence of Hugh McColl, a dour Scot and autocrat who had been allowed to exert undue influence, possibly beyond his competence, especially during Wainwright's declining years. Maunsell had a major task ahead in managing the Works and the design and construction programme, and the directors recognised his priorities and somewhat belatedly split off the responsibility for managing locomotive and rolling stock performance in traffic. His salary was raised to £2,000 for the first twelve months (Wainwright's was only £1,550 for the theoretically larger role) and was raised to £2,500 after that.

Some rebuilding of older classes took place, mainly reboilering, but in 1914 the government created the Railway Executive Committee to take charge of the railways during the wartime period, and Richard Maunsell was appointed as Chief Mechanical Engineer to this body. Some of his work involved the overseeing of maintenance of locomotives in Belgium and Northern France working under ROD auspices and at the end of the war he was awarded the CBE for his services. During this time, however, he still found the time to design his prototype locomotives for the SE&CR, the 'N' class mogul and the 'River' class 2-6-4 express passenger tank engine.

In the period between the end of the war and the Grouping, Maunsell set about rebuilding a number of the Wainwright 'D' and 'E' 4-4-0s with 10in piston valves, long travel gear and superheaters. Twenty-one 'Ds' and eleven 'Es' became 'D1' and 'E1' classes and were so successful that they lasted until the early 1960s. Maunsell had been poised to act as the CME of the proposed nationalised railway after the war, but political views changed and the Grouping proposed by Sir Eric Geddes, then Minister of Transport, came about under the Railways Act of 1921, implemented on 1 January 1923.

With Robert Urie's retirement at age sixty-eight, Maunsell was the natural successor as CME of the new Southern Railway, inheriting a fleet of 2,285 steam engines of 115 different classes, with little standardisation. Richard Maunsell had absorbed some of the best developments from Churchward at Swindon – taper boilers, top

Richard Maunsell at Ashford, c1914. G.M. Rial

feed, Belpaire fireboxes, long-travel, long-lap valve gear – which had already proved their value in the moguls and rebuilt 4-4-0s, then turned his attention to the larger locomotives he'd inherited from Urie on the L&SWR. He went on to develop the N15 and S15 classes, as well as his own 4-cylinder 'Lord Nelsons' and 3-cylinder masterpieces, the V 'Schools' class.

Richard Maunsell was a consummate and skilled manager and administrator, and popular with his team and staff. He gained respect and not a little awe from his noted refusal to tolerate any irregularities however small and was known as 'a terrible straight man'! His influence in new steam engine design was circumscribed by the Southern Board's priority of investment in electrification, restricting money available for the wholesale standardisation of the steam stock as happened on the GWR under Churchward and Collett, and Stanier on the LMS, so older designs were retained in addition to the basic N15, H15 and S15 family which were the nearest the SR got to standardisation in this inter-war period. Maunsell was heavily involved, along with electrical engineers H. Jones and A. Raworth, in the development of rolling stock for the Brighton and Portsmouth electrification, developing the steam-hauled stock he had designed in the mid-1920s. By the time of his retirement, 3,000 coaches of electric stock existed, a tenfold increase from 1923. During this time, the workshops and Maunsell were under great pressure to meet the electrification deadlines and this was achieved despite the stress this caused.

Increasing ill health caused Richard Maunsell to take retirement in 1937, when he was sixty-nine years old. He formally handed over responsibility for the SR Motive Power Department to Oliver Bulleid on 31 October, having had a strong and successful relationship for many years with the Southern's General Manager, Sir Herbert Walker. He left the railway with 1,852 steam engines of 77 classes and a substantially electrified network, retiring to spend his days involved in the life of his local Parish Church, which he and Edith attended regularly in Ashford. He was made an honorary member of the Institute of Mechanical Engineers in 1938 and was often called upon to take organised groups around Ashford Works. His last public appearance was on 7 February 1944 at the Dover Harbour centenary celebrations. He died in March 1944, leaving his wife and married daughter, his only child, and is buried in Bybrook Cemetery, Ashford, a few hundred yards from his house, Northbrooke, where he had lived for thirty-two years.

Chapter 3

THE 'L' DESIGN & CONSTRUCTION

The SE&CR. Class 'L' 4-4-0 was conceived at the end of the Wainwright era and emerged after control had passed to his successor, Richard Maunsell. However, most of the design can be credited to Wainwright's Chief Draughtsman, Robert Surtees, as Wainwright himself was basically a 'carriage & wagon' man and his interest in locomotive design was said to be mainly in the external appearance of his locomotives – for which, it must be said, Wainwright was an artist. That is perhaps too superficial a judgement, for Wainwright was actively trying to come to terms with the increasing train loads in the decade before the First World War, which were taxing his 'D' and 'E' 4-4-0s. Around 1906-7, he had caused a number of designs for heavier passenger engines to be mooted – an atlantic 4-4-2, inside and outside cylinder 4-6-0s and a large boilered inside cylinder 4-4-0. However, civil engineering limitations over the Chatham route to the Kent Coast forced abandonment of these initiatives.

In 1913, he developed an outside cylinder 4-6-0 design with 6ft 6in coupled wheels, 160lb boiler pressure, with 18½ ton axleload, weighing 112 tons (see weight diagram on page 223). The locomotive was to have a Belpaire firebox and Schmidt superheater, but instead a 4-4-0 design was preferred with similar boiler mountings, cab and tender. The civil engineering restrictions on the Chatham route were investigated and a sum of £250,000 to strengthen the route for more powerful engines was considered to be excessive, so Wainwright had to produce a design with continuing axle-load limitations. The first proposed design had 18¾ ton axleload and slide valves, with 180lb boiler and 6ft 6in coupled wheels. A second design with piston valves and 19 ton axle-load was prepared with Belpaire firebox, extended smokebox (like the 'E' class) and Ramsbottom safety valves. Subsequently, Wainwright added a Schmidt superheater as had been proposed for the 4-6-0 and in October 1913, tenders were sought for the construction of up to a dozen engines from Beyer, Peacock as Ashford Works was in no position to build the new locomotives.

In the summer of 1913, locomotive performance was subject to much criticism and Wainwright was held responsible, being persuaded to retire on ill-health grounds. Maunsell was appointed before the contract for the construction of the new 4-4-0 was completed and he and Surtees took the opportunity of reviewing and revising the design. Maunsell took advice from his previous colleagues at Inchicore, particularly the Chief Draughtsman there, W. Joynt, and as well as increasing the coupled wheel diameter to 6ft 8in and extending the boiler length, actually reduced the valve lap from $1\frac{1}{16}$in to $\frac{7}{8}$in which with hindsight might have been a retrograde step. Further detail arrangements were made in extending the cab roof and chimney shape, so that the resultant finalised design was a Wainwright /Surtees/ Maunsell hybrid.

The final settled key dimensions included two inside 20½in x 26in cylinders, 6ft 8in coupled wheels over a 10ft wheelbase, 160lb boiler pressure, 1,412sqft heating surface, Robinson superheater and Belpaire firebox. The engine weighed 56 tons 1 cwt and the 3,450 gallon 4 ton capacity tender weighed 39 tons. The grate area was 22½sqft and the axle-load over the coupled wheels was 18¾ tons. The new locomotives were required for the 1914 summer service to obviate the 1913 experience and the Beyer, Peacock engines, promised for June 1914, only began to appear in August (the first six numbered 760–765) with a further five in September (766-770) and the last, 771, in October. The cost was £4,195 per locomotive (reduced from an initial quotation of £4,385 on the original Wainwright design).

When it became apparent that the Beyer, Peacock contract was running late and Ashford Works did not have the capacity to build the further ten authorised by the Board, Maunsell was able to obtain a contract with Borsig of Berlin to build the locomotives at a cost of £3,790 each, when no further offers were made by British manufacturers. Some cost cutting in materials was made with Maunsell's approval and Schmidt superheaters (which proved more reliable in practice) were substituted. The Borsig engines began delivery in May 1914 (772–778) and were completed in June (779–781) with them being tested by the Borsig fitters and painted in SE&CR Brunswick Green livery with yellow lining and cabside brass numerals, entering traffic the following month. All ten had completed their trial 3,000 mile running by 12 August 1914 and with delivery / completion arrangements slightly modified, cost was reduced to £3,763 – although because of the outbreak of the First World War, payment to the Borsig company was not made until 1920. The axle-load of the Borsig engines over the coupled wheels was slightly heavier at 19 tons 5 cwt and overall weight increased to 57 tons 9 cwt (engine) and 40 tons 6 cwt (tender).

In 1915, Maunsell simplified the smokebox design and fitted a new superheater tested successfully on an Ashford stationary boiler. Trials with 772 against 777 which still had a Schmidt superheater demonstrated a 26 per cent increase in superheat after an hour's working. The Ashford / Maunsell superheater became standard for the class after the Grouping in 1923. New boilers constructed in 1925 included the new superheater and larger firebox water space.

In 1921, in common with railways elsewhere during the first post-war major coal strike, 772 was converted to oil-firing, using the 'Scarab' system. The experiment was not successful and the engine returned to coal-burning within a month after working services between Ashford and Cannon Street and Ashford and Hastings. The experiment had cost £140. The diameter of the cylinders on

Borsig built 779 delivered to the SER in Works Grey before repainting in the company's livery, July 1914. John Scott-Morgan Collection

Borsig built 778 after delivery from Germany and repainted in the SECR company livery, seen here at Ashford, 1914. John Scott-Morgan Collection

The first Ashford built example, 760, at Bricklayers Arms, 1914.
F. Moore/MLS Collection

763 and 766 was reduced from 20½ to 19½in from 1922 to try to improve performance on level stretches of track (such as between Tonbridge and Ashford) where they had seemed sluggish in comparison with the 'Ds' and 'Es' albeit they were much stronger on the banks with the loads then common. In June 1922, an eighty minute schedule had been introduced for the 68 mile run from Charing Cross to Folkestone and the 'Ls' were found to struggle to maintain the schedule east of Tonbridge. It was agreed to convert all the class to this cylinder dimension but to compensate by raising the boiler pressure to 180lbs – further improvements could not be made without a redesign of the front end and valve events, using the experience gained by Maunsell and Holcroft from the GW Churchward developments.

During the First World War, many of the 'Ls' were repainted in the austerity SE&CR grey livery, and in January 1924, 766, renumbered by the Southern Railway as A766, was repainted green, the new company's livery. During the 1926 General Strike, 'L' 763 was named *Betty Baldwin* by one of the volunteer crews (a girlfriend of one of the drivers perhaps?) and bore this name on the leading splashers until May 1927.

Many of the improvements – cylinder diameter reduction,

A763 was named *Betty Baldwin* during the 1926 General Strike, at Bricklayers Arms, c1926.
J.M. Bentley Collection

boiler pressure increase, Maunsell superheater – were only implemented gradually as 'Ls' were no longer on the most prestigious services, and the cylinder changes continued to be implemented throughout the 1930s with the last, 780, not converted until April 1941. Similarly, the Maunsell superheater exchange was not completed until November 1938 (778). The increase in boiler pressure, although carried out at the same time as the cylinder diameter reduction on some locomotives, was not completed until September 1944 (780). With boiler exchanges, however, some engines that had been already modified reverted to older conditions and vice versa. In 1931, the Southern Railway dispensed with the A, B and E prefixes to the engine numbers and the Eastern Section locomotives had 1000 added, so that A760 became 1760, etc.

At the commencement of the Second World War the 'Ls' were repainted a plain dark green, then in 1942 black was substituted for green. During the war, lines

1761 at Cannon Street as repainted and renumbered by the Southern Railway in 1931. This locomotive is the sole 'L' that received modified valve arrangements giving longer travel and lap and was subsequently reputed to be more free-running than the others of its class. Real Photographs/ John Scott-Morgan Collection

31769 at St Leonards depot, Hastings, repainted in malachite green after the Second World War, renumbered to its new BR notation, but still inscribed 'Southern' on the tender, c1949. J.M. Bentley Collection

1764, repainted malachite green after the Second World War, seen here at Stewarts Lane, 11 August 1947. MLS Collection

31768 at St Leonards, repainted BR mixed traffic lined black, but before the lion & wheel motif was approved, with the 'C' class tender that this engine retained for many years following the 1937 Swanley Junction accident, 23 July 1953. MLS Collection

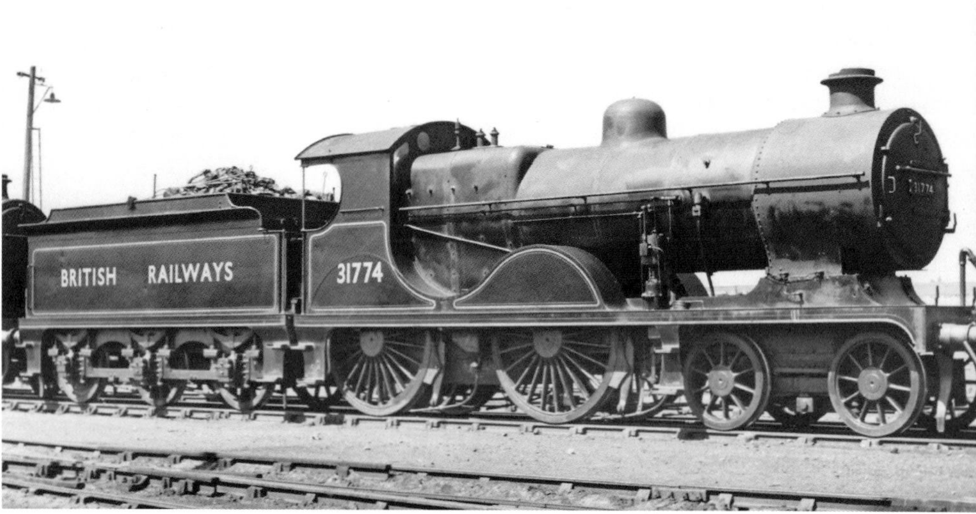

31774 repainted in BR mixed traffic livery and 'British Railways' inscribed on the tender before the 'lion & wheel' motif was utilised, photographed at Ashford, 9 July 1950. W. Gilburt/John Scott-Morgan

Borsig 31781, somewhat careworn, at Ramsgate shed, 12 July 1954. MLS Collection

to the Kent Coast were busy and mileages between general repairs were increased, with 1772 achieving 111,268, the most recorded. The black livery continued until June 1946, when it was replaced by malachite green. All bar six (1760/2/3/5/71/6) received this livery before the BR mixed traffic lined black became standard. The last 'L' wearing the malachite green livery was 31780, repainted mixed traffic black in July 1954.

During the 1950s, many 'Ls' were put into store and after 1957 few general repairs and repainting took place. The first to be withdrawn were 31761

A rear view of 31766 in BR mixed traffic livery with the small BR 'lion & wheel' motif, at Eastbourne shed, 7 October 1950.
H.C. Casserley/J.M. Bentley Collection

One of the last 'Ls' to remain in traffic, the Borsig-built 31776, in BR mixed traffic livery with the late BR motif on the tender, seen here at Newhaven, 13 April 1958.
G. Shuttleworth/MLS Collection

and 31769 in 1956 with mileages of 1.26m and 1.12m miles respectively. 31767 and 31774 were condemned in 1958 and eight were withdrawn in 1959. The Kent Coast electrification in June 1959 made most of the remaining members of the class redundant, with just four (31762/71/76/77) still active. In 1960, consideration was given to the preservation of 31763, but no funds were available and it was scrapped in 1961. 31768 was resuscitated to join 31776 and worked in the SR's Western Section, the last to survive, finally being withdrawn in December 1961 having achieved a mileage of 1,568,447 in its lifetime.

Operation

The 'Ls' entered service mainly on three-coach stopping trains between Ashford and Tonbridge, Hastings and Margate during their initial trial period. At the end of this time, they were allocated to:

Bricklayers Arms: 760, 762, 763, 770, 772, 773, 774
Dover: 761, 775, 777, 778, 779
Hastings: 764, 766, 769, 771
Ashford: 776, 780
Cannon Street: 765, 781
Ramsgate: 767, 768

They then took over the most difficult Eastern Section expresses, especially the Dover and Folkestone boat trains. They performed excellently on the banks and the crews valued their smooth riding and free steaming boilers. Some track problems around Hastings caused a few minor derailments and in 1915, the Hastings engines were reallocated to Bricklayers Arms whilst the track defects were sorted out. Maunsell was pleased with their apparent low coal consumption (averaging around 33–34lb per train mile) and used the 'L' boiler for his 2-6-4 freight tanks (the 'W' class) and his moguls. Mileages between heavy repairs ranged from 60,000 to 84,500, the average over a three year period in the First World War being 71,250.

Although built for acceleration of heavy continental and Kent Coast expresses, the 'Ls' found considerable use during the war on slow speed heavy troop and ambulance trains. The SE&CR routes were crucial to the war effort and the delivery of manpower and ammunition to the Channel ports. However, the collapse of part of the steep cliffs between Dover and

Borsig-built 773 at London Bridge with a train of four-wheeled rolling stock, 1914.
J.M. Bentley Collection

771, the last Ashford built 'L', on a London Bridge–Hastings train, 1914. J.M. Bentley Collection

770 in wartime grey livery, at Grove Park with a train of refurbished suburban six-wheel rolling stock, 1920. Photomatic/John Scott-Morgan Collection

Folkestone in 1915 blocked the route over which the 'Ls' worked and for a while the company was forced to rely on the 'Ds' and 'Es' and Stirling's 'B' and 'F' 4-4-0s to take everything over the congested and more difficult LC&DR route from which the 'Ls' were banned because of their weight.

After the war, services were accelerated and in 1922 eighty-minute schedules from London to Folkestone were introduced. Loads were around 220 tons including a Pullman Car and drivers began to experience problems in maintaining sufficiently high speeds over the Tonbridge–Ashford 'racing' section. The comparative sluggishness was attributed to the cylinders beating the boiler, and, as mentioned earlier, proposals were developed to reduce the cylinder diameter and increase the boiler pressure to address the problem. The real solution, however, was found in the building of fifteen 'improved Ls' – the 'L1s' (see chapter 6). In the interim period, the shortcomings of the 'Ls' were addressed by transferring ten Drummond 'L12' 4-4-0s from the SR's Western Section, although their improved performance on the road was to some extent nullified by their propensity to suffer hot axle-boxes, a problem that was not suffered by the 'Ls'.

Cecil J. Allen published an interesting run in 1925 on the 3.35pm Charing Cross–Hastings as far as Crowhurst, recouping all of a four minute late departure and a bit more with a nine coach 270 gross ton train, hauled by 'L' 770 of Bricklayers Arms. The engine working south of Tunbridge Wells was particularly impressive.

Charing Cross–Crowhurst, Summer 1925
3.35pm Charing Cross–Hastings
A770 Bricklayers Arms
9 chs 251/270 tons

Miles	Location	Times Min secs	Speed mph	Punctuality
0.0	Charing Cross	00.00		3½ L
1.8	London Bridge	04.40	25/52	3¼ L
4.9	New Cross	09.50	sigs	4 L
7.1	Hither Green	13.10	43½	4¾ L
8.9	Grove Park	15.50	36	
10.2	Elmstead Woods	18.20	33	
11.2	Chislehurst	19.55	40	5 L
13.8	Orpington	23.35	46	5 L
15.2	Chelsfield	25.40	40	
16.5	Knockholt	27.35	37	5 L
20.5	Dunton Green	32.05	68	
22.0	Sevenoaks	32.35	54	4½ L
26.9	Hildenborough	38.15	72½	
29.5	Tonbridge	40.55	40*	4 L
32.9	High Brooms	46.55	31½	
34.4	Tunbridge Wells	49.35	40*	3½ L
36.7	Frant	52.55	36½/61½	
39.3	Wadhurst	55.45	53/60*	3¼ L
43.8	Ticehurst Road	60.20	65	
47.4	Etchingham	63.30	74	
49.6	Robertsbridge	65.35	45*	1 L
	Mountfield sidings	-	61½	
55.5	Battle	72.15	47½	¼ E
57.6	Crowhurst	75.30	(74 net)	1 E

A776 on a Victoria–Ramsgate via Chatham train, c1923. The second, third and fourth vehicles are steel bodied 100-seat coaches built originally for suburban electric stock. F.E. Mackay/ J.M. Bentley Collection

A772 and another 'L' standing at the head of a train at Charing Cross, the headcodes not yet put in position to denote its destination, c1925. John Scott-Morgan Collection

A774 on a down excursion for Deal, composed mainly of former LB&SCR rolling stock, at Elmstead Woods, c1925. J.M. Bentley Collection

A764 on an up express at Shorncliffe, July 1931.
J.M. Bentley Collection

During the 1926 General Strike with the few heavy trains calling frequently at intermediate stations, the 'Ls' came into their own and with the 'N' 2-6-0s dominated the services which ran. With the introduction of the N15 'King Arthurs' in 1925-6 and the removal of the Chatham line weight restrictions, A767, 768 and 771 were transferred to Battersea (Stewarts Lane) and A774, 778 and 779 to Ashford. They remained the principal motive power for the Charing Cross–Hastings services. Then in 1930 came the introduction of the 'Schools' and the ''Ls' lost their monopoly and the allocation was again adjusted as follows:-

Hastings: A760-766
Battersea: A767-771
Ashford: A772-776
Dover: A777-781.

The most demanding diagram for an 'L' after that time was for one of the Ashford engines which was booked:-

6.28am Ashford – Victoria via Maidstone East
10.35am Victoria – Ramsgate
4.25pm Ramsgate – Victoria
9.33pmVictoria – Ashford

Most of their work was on the Eastern Section though occasional incursions into Brighton territory on excursion or summer holiday traffic occurred from time to time. From June 1934 the 'Ls' then allocated to the Battersea Stewarts Lane depot would work regularly to Eastbourne and Bognor until those routes were electrified. I have found no more records of running of 'Ls' on the faster services they hauled in SE&CR days or early days of the Southern Railway, but a regular traveller in Kent in the 1930s, 40s and early 50s, Mr. S.A.W. Harvey, recorded a large number of logs found in the records of the Railway Performance Society and I have made a random selection to represent their running. The earliest I have found dates from the summer of 1933.

However, before giving details of the logs, I'll describe the gradients over the two main routes. The route from Charing Cross, Cannon Street or London Bridge to Folkestone and Dover via Tonbridge and Ashford is level until Hither Green, then rises for over three miles at 1 in 140/120 to Elmstead Woods and after a short level stretch, rises at 1 in 234/310 to Orpington steepening to three miles of 1 in 120 to Knockholt summit. It's then falling at 1 in 143 through Polhill Tunnel to Dunton Green (4 miles) with a mile and a half of 1 in 160 up to Sevenoaks. There is then a long (6 miles) descent at 1 in 144/122 to Tonbridge. After Tonbridge, there is a slight undulation to Paddock Wood (1 in 270 up/down) a slight rise to Marden and then fall again, then level to Headcorn and then a steady rise to Ashford varying 1 in 250/280. After Ashford there is a long (8 miles) climb at 1 in 250/280 to Westenhanger, then a drop just before Sandling Junction at around 1 in 264 all the way to Dover (about 11 miles). If the train starts from Victoria, it has to face a sharp 1 in 62 up to the Grosvenor Road Bridge over the Thames.

The former LC&DR route via Chatham, after the start from Victoria, faces a 1 in 100 climb from Herne Hill to Sydenham Hill (2 miles), a drop after Penge Tunnel at similar gradient to Beckenham Junction, a climb from Shortlands to Bickley Junction (nearly three miles 1 in 95), a mile drop at 1 in 100 through St Mary Cray then undulating to Swanley, three miles falling at 1 in 100 to Farningham Road (where maximum speeds in both directions are usually recorded), two miles up at 1 in 100 to Fawkham Junction, another two miles through Meopham to Sole Street at 1 in 100/132, then a five mile drop at 1 in 100 from Sole Street to Rochester Bridge through Cuxton Road. After Chatham, there is a mile rise at 1 in 135 to Gillingham, then level through Rainham and Newington and after a short rise, a fall at 1 in 100/120 to Sittingbourne. Undulating grades through Teynham, 1 in 132 up and down to Faversham, then largely level (slight undulations) to Herne Bay, then two miles 1 in 93/100 up and three miles similarly down to Reculver. After Birchington, a mile of 1 in 100 down, then undulations to Margate.

Gradient profiles of both routes can be found in the Appendix.

Charing Cross–Margate, 3.7.1933
9.15am Charing Cross-Margate
1777, Driver Reynolds
256/286 tons
(Timed by S.A.W. Harvey)

Miles	Location	Times	Speed	Punctuality
0.0	Charing Cross	00.00		T
0.8	Waterloo East	02.30	30	½ L
1.9	London Bridge	06.50	sigs stand ½ min	
4.9	New Cross	13.10	48	¾ L
5.6	St Johns	14.10	52	
7.2	Hither Green	16.30	46	½ L
10.3	Elmstead Woods	21.57	36	
11.3	Chislehurst	23.35	44	½ L
13.8	Orpington	27.18	45/50	1¼ L
15.3	Chelsfield	29.15	46	
16.6	Knockholt	31.05	45	
20.6	Dunton Green	35.27	70	
22.1	Sevenoaks	36.50	58*/68	¾ L
27.0	Hildenborough	41.28	78	
29.5	Tonbridge	44.55	(40.45 net)	T
0.0		00.00		T
5.3	Paddock Wood	07.42	64	¼ E
9.9	Marden	12.25	62	
12.4	Staplehurst	14.55	64	T
15.7	Headcorn	18.22	66	
20.9	Pluckley	24.15	56	
26.6	Ashford	31.18		¾ E (¼ L WTT)
0.0		00.00		T
4.3	Smeeth	07.30	50	
8.1	Westenhanger	11.55	56	
9.3	Sandling Junction	13.22	60	1¼ L
12.0	Cheriton	17.10	pws/40*	
13.1	Shorncliffe	18.30	48/56	
13.8	Folkestone Central	19.40	(18.00 net)	¾ L (2 ¾ L WTT)

The train then called at all stations via Dover Priory, Deal and Sandwich, arriving early by the public timetable and just about on time by the WTT. However, the 'L' came off the train at Deal and was replaced by an 'H' tank for the last short stretch to Margate. The same engine figured in an up run a couple of years later and I show below the most interesting part of the journey from Chatham into London.

Chatham–Bromley South, 25.5.1935
7.12pm Chatham-Victoria
1777, 250/270 tons
(Timed by S.A.W. Harvey)

Miles	Location	Times	Speed	Punctuality
0.0	Chatham	00.00		T
0.6	Rochester	01.30	30	
1.4	Rochester Bridge	02.55	40	½ L
3.4	Cuxton Road	07.45	30	
7.4	Sole Street	16.50	30	¾ L
8.4	Meopham	18.30	58	
10.9	Fawkham	21.12	78	
13.8	Farningham Road	23.37	80	
16.6	Swanley Junction	26.25	sigs 55*	½ E
20.9	St Mary Cray Junction	32.00	65/50*	
21.7	Bickley Junction	33.00	50*/55	T
23.4	Bromley South	35.58	(33.45 net)	T

The train then continued to Victoria, maximum 60mph at Beckenham Hill, and after a severe pws at Denmark Hill, arrived at Victoria ¾ minutes late. The 'Ls' had the reputation of being less free-running than the later 4-4-0s (to some extent blamed on the retrograde step taken by Maunsell and Joynt of reducing the valve travel and lap), but not only was 80mph achieved here at a traditional high speed location, but Mr Harvey also provided a most surprising snippet with 1774 when – admittedly with a light four-coach train only – it accelerated the 8.11am Ramsgate–Ashford from Grove Ferry on the Minster–Canterbury West route from 45mph to 75mph in just two and a quarter minutes on a slightly rising gradient. With a top speed of 76mph and another 74 at Wye after the Canterbury stop, it was a very enterprising little run.

1777 on a down Folkestone Pullman car train, passing Ashford, 30 September 1933. Pamlin print/J.M. Bentley Collection

1778 at Knockholt with a down express, 6 January 1934. H.C. Casserley/ J.M. Bentley Collection

1780 on a down Hastings train at Mountfield, c1935. J.M. Bentley Collection

Next, we have a run via the LC&DR route to Ramsgate in the summer before the Second World War:

Victoria–Ramsgate, 30.6.1939
10.10am Victoria–Ramsgate
1763, Driver Bullock
315/325 tons
(Timed by S.A.W. Harvey)

Miles	Location	Times	Speed	Punctuality
0.0	Victoria	00.00		T
0.7	Grosvenor Road Bridge	02.10		
2.3	Clapham	05.40	46/40*	
3.2	Brixton	07.12	40	
4.3	Denmark Hill	08.55	62	
6.0	Nunhead	11.25	sigs	1½ E
7.8	Lewisham Junction	16.04	sigs	T
0.0		00.00		T
1.3	Hither Green	03.25		1½ E
4.4	Elmstead Woods	09.30	36	
5.4	Chislehurst	11.10	40	1¾ E
7.8	St Mary Cray	14.40	62	
10.6	Swanley	17.58	pws	2 E
13.5	Farningham Road	21.38	76	
16.4	Fawkham	24.30	60 eased	
18.9	Meopham	27.40	56	
19.9	Sole Street	28.55		1 E
23.9	Cuxton Road	32.45	78	2¼ E
25.9	Rochester Bridge	35.22		3¼ E
27.3	Chatham	38.35	sigs	2½ E
28.9	Gillingham	43.20	sigs	
31.9	Rainham	48.00	55	
34.6	Newington	51.02	58	
37.7	Sittingbourne	56.05	sigs	3 L
41.0	Teynham	60.07	66	
45.0	Faversham	64.36	45*	3½ L
48.1	Graveney	68.05	66	
52.1	Whitstable	72.34	(64.15 net)	2½ L

The train then continued all stations to Ramsgate holding schedule but unable to regain any time because of a further severe pws outside Burchington.

1768 was involved in a serious accident in June 1937. The late running 8.17am Margate–Victoria train was to stop additionally at Swanley Junction to pick up passengers from an earlier missed connection, but the driver misread the junction signals and crashed into loaded wagons and a motor set at 40mph, killing four passengers and severely injuring eleven. The engine was repaired but the damaged tender was scrapped and replaced by a smaller 'C' class tender which it kept until January 1957, when the tender from the withdrawn 31765 became spare.

In the mid-late 1930s, despite the availability of more powerful and speedier classes, the 'Ls' retained some express work, including some boat trains, Charing Cross–Hastings trains and regular piloting of the *Night Ferry* introduced in 1936. Another regular diagram included the Redhill-Kent Coast portion on the through train from Birkenhead, nicknamed for many years as the 'Conti'. In May 1939, four 'Ls' (1765/66/70/78) were reallocated to New Cross replacing a similar number of un-rebuilt 'E' 4-4-0s, and worked, among other Central Section services, the important commuter 5.08pm London Bridge–Three Bridges–Forest Row train. Other work included excursion and school traffic to the Sussex Coast and sorties into the SR's Western Section. The allocation at the beginning of the Second World War stood as follows:

Battersea (Stewarts Lane): 1762-1764
New Cross Gate: 1765, 1766, 1770, 1778
Dover: 1760, 1761
St Leonards (Hastings): 1767, 1768
Ashford: 1769, 1771-1777
Bricklayers Arms: 1779-1781

Initially several 'Ls' were put to store, but in January 1940 all returned to traffic and were heavily involved in wartime troop

1774 at Sandling Junction, April 1938.
J.M. Bentley Collection

1773 at Sevenoaks with a Hastings train, passing ex LBSCR D3 0-4-4T 2367 on a push-pull train to Tonbridge, c1946.
John Scott-Morgan Collection

movements, especially the army evacuation from Dunkirk (as many as seventeen 'Ls' were noted in this work, including heading the ambulance trains). 1766 and 1767 were transferred to Ashford and 1770 and the three Bricklayers Arms engines moved to Tonbridge. Working almost exclusively on Kent Coast services, the 'Ls' were particularly vulnerable to enemy air attack and 1778 narrowly evaded bombing near Paddock Wood whilst working a Tonbridge–Ashford stopping train. During the war a number of Western Section 4-6-0s (Urie 'Arthurs', 'H15s' and 'N15Xs') were loaned to other railway companies for heavy freight work, and Eastern Section engines filled some of the gaps with 'Ls' covering the resultant secondary 'vacancies'. 1775–1779 went to Faversham and Ramsgate in January 1943 and then three of them (1775–1777) moved on in February 1944, along with five 'King Arthurs' to Ashford. Time between heavy repairs was extended during the war years, with 1772 running over 111,000 miles before Works attention.

After the war, the revised allocation was:

Stewarts Lane: 1760-1765
St Leonards: 1766-1769
Ashford: 1770-1777
Ramsgate: 1778-1781

The St Leonards engines were regularly rostered to the 10am Hastings-Charing Cross and 6.03pm Cannon Street return, both relatively heavy trains (8-9 coaches) and the Stewarts Lane engines worked to both Eastern and Central Section destinations including Newhaven boat train work, especially when double-heading was required at peak times.

After nationalisation, little changed until the advent of the LMS 2-6-4 tanks built at Brighton which reached the area in force in 1951, replacing the 'I3' tanks on the Brighton line and removing much of the work that the 'Ls' had covered in Kent. Experimentally, 31770 was transferred to Eastleigh in 1951 as a possible replacement for older Drummond 4-4-0s (such as the 'L11s' and 'T9s') and started work on a diagram including stopping services to Bournemouth and Andover Junction. It was received well enough for a further ten (31770-31779) to be transferred in January 1952 and they were employed on a variety of stopping passenger and local goods trains, including the Bournemouth-Salisbury services, the Botley area strawberry trains and goods trains on the Alton branch. However, another influx of LMS and BR standard tank engines replaced them once more and four (31770-31773) returned to Tonbridge in August 1952. Then the remainder were replaced at Eastleigh by 'L1s' and departed for Ashford.

A selection of post-war logs from the Railway Performance Society archives is now shown below, starting with two late night passenger & mail trains which now featured regularly among the 'Ls' duties:

London Bridge-Sevenoaks

		10.25pm Charing Cross-Wadhurst, 27.12.1950			**10.50pm London Bridge-Wadhurst/Dover**		
		31779			**31766**		
		4 chs + 1 van, 135/150 tons			**7 chs + 1 van, 234/245 tons**		
		(Timed by B.I. Nathan)			**(Timed by B.I. Nathan)**		
Miles	**Location**	**Times**	**Speed**	**Punctuality**	**Times**	**Speed**	**Punctuality**
0.0	London Bridge	00.00		T	00.00		T
3.0	New Cross	05.57	30½		06.22 pws 30*		
3.7	St Johns	06.51			07.17	45½	
5.3	Hither Green	09.07	43/40		09.27	43/39	
8.4	Elmstead Woods	14.00	36½		14.05	38½	
9.4	Chislehurst	15.26	42		15.30	41/43	
11.9	Orpington	18.47	52		18.59	51/47½	
13.4	Chelsfield	20.28	53½		20.48	44	
14.7	Knockholt	22.09	46½		22.31	42½	
18.7	Dunton Green	26.31	55/sigs		26.37	67	
20.2	Sevenoaks	29.09	(28.15 net)	T	29.06	(28.30 net)	T

The second run in particular with the heavier load (the train split at Tonbridge) was excellent – timekeeping was important on this train with the mails and newspapers.

Margate-Bromley South, 7.6.1949
7.55pm Margate-Victoria
31781
255/270 tons
(Timed by S.A.W. Harvey)

Miles	Location	Times	Speed	Punctuality
0.0	Margate	00.00		T
1.5	Westgate	04.01		
3.2	Birchington	07.08		
7.2	Reculver	12.00	54	
9.6	MP 64 ¼ (summit)	16.15	38/30	
11.2	Herne Bay	19.25		2½ L
		00.00		
2.2	Chestfield Halt	04.00		
3.6	Whitstable	06.37		2 L
		00.00		
4.0	Graveney Sdgs	06.34	58	
7.1	Faversham	10.59		3 L
		00.00		
4.0	Teynham	06.48	70	
7.3	Sittingbourne	11.04		3 L
		00.00		
3.1	Newington	07.03	58	
5.8	Rainham	10.20	60	
8.8	Gillingham	15.44	sigs (14.45 net)	5¾ L
		00.00		
1.6	Chatham	03.57		5¾ L
		00.00		
1.4	Rochester Bridge	02.50		6 L
3.4	Cuxton Road	06.30	36	
7.4	Sole Street	15.02	28	4 L
10.9	Fawkham	19.50	70	
13.8	Farningham Road	23.15	sigs	
16.9	Swanley	29.12		6¼ L
19.4	St Mary Cray	32.35	66	
21.7	Bickley Junction	35.18		6¼ L
23.4	Bromley South	38.05	(34.15 net)	5¾ L

31781 was struggling to keep time, its strongest performance being on Sole Street bank, and just when it was promising to regain substantial time, it caught signals on at Farningham Road, the fastest section.

Hastings-Charing Cross, 11.9.1950
11.15am Hastings-Charing Cross
31766
256/280 tons
(Log by S.A.W. Harvey)

Miles	Location	Times	Speed	Punctuality
0.0	Hastings	00.00		T
0.7	Warrior Square	02.56		T
		00.00		
1.0	West St Leonards	02.51		¼ E
		00.00		
3.1	Crowhurst	08.10		2 L
		00.00		
2.0	Battle	04.52		3 L
4.9	Mountfield	08.10		
8.0	Robertsbridge	11.40		
10.2	Etchingham	14.00	46	2 L
18.4	Wadhurst	26.44		3¾ L
23.3	Tunbridge Wells Central	34.33		3½ L
		00.00		
1.5	High Brooms	04.05		
4.9	Tonbridge	08.07		3½ L
7.4	Hildenborough	12.40	38	
11.7	Sevenoaks Tunnel (N)	20.08	39	
12.3	Sevenoaks	21.23		2¾ L
		00.00		
1.5	Dunton Green	02.51		
5.5	Knockholt	08.10	50	
6.8	Chelsfield	09.35	60	
8.3	Orpington	11.02	68	2¾ L
10.8	Chislehurst	13.30	66	2¼ L
11.8	Elmstead Woods	14.23	68/70	
14.9	Hither Green	17.03	70	1¾ L
17.2	New Cross	19.38		1¼ L
20.2	London Bridge	23.45	sigs	1 L
21.3	Waterloo East	26.45		½ L
		00.00		
0.8	Charing Cross	03.15		¾ L

A typical run by an 'L' on a Hastings train, losing a little time on the stopping sections but climbing well out of Tonbridge and Sevenoaks and coasting downhill at 70mph.

Charing Cross-Ashford, 17.7.1954
2.15pm (SO) Charing Cross-Dover Priory
31776
329/355 tons
Weather – pouring rain all the way

Miles	Location	Times	Speed	Punctuality
0.0	Charing Cross	00.00		T
0.8	Waterloo East	02.18	sigs	
1.9	London Bridge	06.42	sigs	¾ L
4.9	New Cross	11.44		1¾ L (blowing off steam)
7.2	Hither Green	15.18	46	2¼ L
10.3	Elmstead Woods	20.24	41/38	
11.3	Chislehurst	21.59	44/sigs	1 L (blowing off)
13.8	Orpington	26.39	45*	1½ L
15.3	Chelsfield	28.50	44	
16.6	Knockholt	31.22	sigs	
20.6	Dunton Green	37.17	65	
22.1	Sevenoaks	39.10	sigs (severe)	3¼ L
27.0	Hildenborough	46.35	80	
29.5	Tonbridge	49.13	30*	5¼ L
34.8	Paddock Wood	55.20	70	5½ L
39.4	Marden	59.40	pws 25*	
41.9	Staplehurst	63.25	60	
45.2	Headcorn	66.44	65	
50.4	Pluckley	71.40	68/60	
56.1	Ashford	79.29	(65.30 net)	6½ L

31770 on the 9.55am Eastleigh-Bournemouth at Swaythling, 19 January 1952.
L. Elsey/J.M. Bentley Collection

This was an exceptional run for an 'L' with this load on a summer Saturday, plenty of steam, and too energetic for the train in front.

Most of the class were unemployed in winter and were placed in store, although maintained in good working order. St Leonards still used its three (31767-31769) on a morning Hastings-Charing Cross train, returning on the 3.25pm Charing Cross-Wadhurst and then locals in the Tonbridge area. In 1956 31776-31778 were transferred to Brighton for Tunbridge Wells West-Brighton services, as well as the occasional special, such as when 31776 piloted a 'Schools' on heavy tour trains from France to the North of England via Willesden at Easter 1958. 'Ls' also started appearing on the Brighton-Bournemouth services, replacing the 'H2' Marsh atlantics. After 1957 few general repairs were carried out. Two 'Ls' had already been withdrawn and 31767 and 31774 were condemned in 1958.

The Hastings route was dieselised in June 1958, making the three St Leonards 'Ls' redundant. The Kent Coast electrification in June 1959 dealt a final blow to their work in Kent and Faversham's 31765/66/68 and Ramsgate's 31764/75/79/80 were stored at Brighton and Nine Elms. 31762 and 31771 remained occupied with Redhill-Brighton and Eastbourne services (such as the portions of the 'Conti') whilst Brighton's 31776 and 31777 worked parcels trains and locals to Tonbridge. The Tonbridge-Redhill-Guildford line also saw them on three coach all stations trains.

The last active 'Ls' in 1960 were 31768 and 31776 working from Nine Elms on empty stock from Clapham Junction, parcels trains to Reading and the Saturday 12.42pm Waterloo-Basingstoke semi-fast. 31768 performed on a railtour from Waterloo to Salisbury in August (see log on page 39) and worked Christmas parcels trains to Richmond and Kingston in December. A few similar workings remained in 1961,

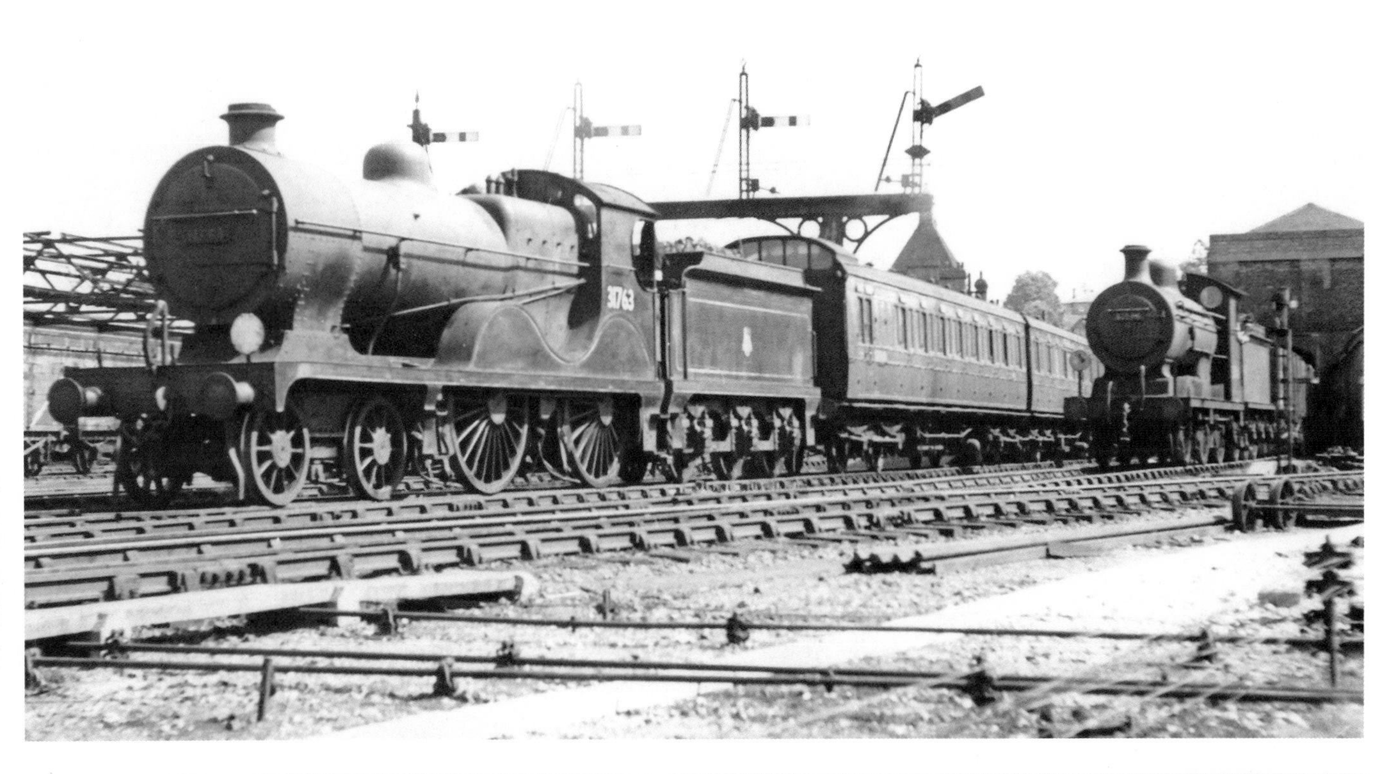

31763 at Tunbridge Wells passing a C2X 0-6-0 in the yard, 3 September 1955.
J.M. Bentley Collection

31767 on an Ashford-Hastings train, possibly near Winchelsea Halt, 1 September 1955.
J.M. Bentley Collection

31781 at Dover Marine, 4 August 1955. L. Hanson/J.M. Bentley Collection

31779 heads a relief to the Ramsgate-Birkenhead 'Conti' at Redhill, 4 August 1955. B.K. Green/MLS Collection

31774 enters Brixton with a semi-fast service, 7 August 1955.
J.M. Bentley Collection

31760 at Gatwick Airport with an up service, c1956.
J.M. Bentley Collection

31779 departing from West St Leonards station with a stopping train for Tonbridge, 11 September 1956. H. Townley/ J.M. Bentley Collection

31776 leaving Brighton with a parcels train, c1958. J.M. Bentley Collection

31776 succumbing in February and 31780 and 31771 re-emerging while those in store were condemned and towed away for the breaker's yard. 31768, having achieved the class highest mileage of 1,568,447, with 31771, were the last to be withdrawn in December 1961.

The 'L', nearing withdrawal, just about met the relatively fast schedule but could not recover from the Sunday engineering heavy delays out to Clapham Junction. The train was taken forward by a 'T9' (hence the title 'Greyhound Special'). The return run timed by both B.I. Nathan and S.A.W.

Waterloo-Salisbury RCTS *Greyhound Railtour*, Sunday, 14.8.1960
10am Waterloo-Salisbury-Weymouth
31768 (Nine Elms) to Salisbury
6 chs, 196/210 tons
(Timed by B.I. Nathan)

Location	Times	Speed	Punctuality
Waterloo	00.00		¼ E
Vauxhall	04.58	sigs 15*	
Clapham Junction	11.00	34/pws 18*	3¾ L
Wimbledon	18.19	50	
New Malden	21.34	57	
Surbiton	24.00	58½	
Hampton Court Junction	25.18	65	7¼ L
Hersham	27.48	64	
Walton-on-Thames	-	65	
Weybridge	30.50	61	
West Weybridge	-	69	
West Byfleet	34.10	69	
Woking	35.42	59	6½ L
MP 31	42.37	57½	
Farnborough	44.52	65	
Fleet	47.56	67	
Hook	53.36	71	
Winchfield	-	66	
Basingstoke	58.06	60/sigs 48*	
Worting Junction	61.00	52/61	7¾ L
Oakley	63.29	56/63	
Overton	66.41	67	
Whitchurch	70.04	65/67	
Hurstbourne	71.46	70	
Andover Junction	76.31	59	
Grateley	83.07	53/67	
Porton	88.21	75/48*	
Tunnel Junction	92.23	15*	8¾ L
Salisbury	94.26	(87 net)	8¼ L

Harvey, was disappointing, taking 102½ minutes, 94 minutes net, with a very slow start whilst the driver tried to remove grit from his eye, then a laboured ascent to Grateley starting from the slow speed (45mph maximum), falling to 38 at Overton, nothing over 60 before Worting Junction, passed nearly 8 minutes late, maximum of 70mph at Fleet, falling to 52 at MP 31, 11½ minutes late at Woking passed at 62mph only, no more than 64mph en route to Hampton Court Junction, a heavy pws at Earlsfield, Clapham Junction passed 15 minutes late and arrival 15½ late.

31781 at Dover Priory station, c1959. J.M. Bentley Collection

Personal Reminiscences of the 'Ls'

I was a 'South Western' man, school days at Surbiton and Godalming, commuting to London University from Woking to Waterloo. After the Kent Coast electrification in 1959, I became very familiar with the 'Schools' (see Chapter 7) but the former SE&CR 4-4-0s and their rebuilds were strangers to my experience. Whilst a few 'E1s', 'D1s' and 'L1s' strayed to the Western Section towards the end of their lives, I'm not aware that any the Wainwright/Maunsell 'Ls' ever did, at least not in their last days. However, I was familiar with the Guildford-Reading section of the Tonbridge-Redhill-Guildford-Reading route, colloquially known as the 'Rattler' (for fairly obvious reasons). Through the 1950s, the Southern Region motive power management tended to see the line as a retirement home for tired locomotives, the most modern steeds being the capable Maunsell moguls of both 'N' and 'U' variety, though in the middle of the decade the occasional Standard 2-6-4T surprised us by its presence. Then in 1955, Redhill acquired ten brand new Standard 4 2-6-0s (76053-62) and swept the pre-grouping designs away within the year.

In their last throes, I came across a couple of 'Ls' and managed to get short runs behind them, though it was in the days before I became an enthusiastic recorder of locomotive performance, my experience limited to being an avid reader of Cecil J. Allen's monthly articles in *Trains Illustrated.* The first was from Wanborough, the nearest station on a non-electrified route to my school, and I rode behind the former *Betty Baldwin*, 31763, in the Maunsell 3-coach set plus parcels van to the next station, Guildford, just five miles away. I did have my camera with me and rushed over to the adjacent platform to snatch a quick shot before it departed for Redhill, although I suspect my haste was unnecessary as these trains pursued a leisurely course through the beautiful Surrey hills.

Tonbridge's 31763 at Guildford on a Reading-Redhill train in which I rode, waiting here at Guildford ready for departure, 1955.
David Maidment

A few weeks later, the Charterhouse Railway Society had an outing to Tonbridge and a shed visit and our train back to Guildford was hauled as far as Redhill by 'L' 31777, seen here at Tonbridge before departure. This engine was, of course, one of the 'Germans' built by Borsig in 1914. Again, I took little interest in the running, although it's possible that a couple of fellow club members, already keen train timers, might have taken a rudimentary log of the running. We stopped at all stations on our three coach local and I doubt if we reached 50mph between stops. At Redhill a new Standard '4' 2-6-0, 76054, backed on and took us to Guildford.

31777 at Tonbridge before departure on the stopping train to Redhill and Guildford on which other Charterhouse Railway Club members and I returned, 1955.
David Maidment

My final encounter with an 'L' was a surprise. By 1958, I was a student at London University studying German language and literature. The source of the cheapest books we had to study was a bookshop close to Charing Cross station, and occasionally I would buy a return ticket from Waterloo East to Charing Cross to get a rumble over Hungerford Bridge behind a St Leonards 'Schools' or a Ramsgate 'Battle of Britain'. On one occasion, having bought my Schiller play (East German paperback version 9d), I returned intending to go back to Waterloo and see what was working the 12.54pm Waterloo-Salisbury semi-fast. There was an 11.46 Charing Cross-Ashford via Otford, a few coaches and a motley collection of bogie vans, usually hauled by a 'Schools' or, if I was lucky, an Ashford six-wheel tender 'King Arthur'. I wandered up the steps onto the footpath that ran across Hungerford Bridge to the South Bank site and could see a 'Schools' at the head end, but another engine coupled ahead of it. I could discern the outline of an 'L' and rushed to the booking office to purchase a single to London Bridge – the most I could afford on this occasion. When I got to the front of the train, I found the pioneer 'L', Beyer, Peacock's 31760, ahead of 'Schools' 30929 *Malvern*. I was again lucky enough to have my camera with me and captured a photo of the pair, before the amble across the Thames and the threading of Borough Market Junction – ten minutes of serenity without breaking the 30mph speed limit, I guess! I presume the 'L' was being worked back to Tonbridge or Ashford this way rather than occupying a path as a light engine on such a congested route.

31760 pilots 30929 *Malvern* on the 11.46am Charing Cross-Ashford via Otford, 1958.
David Maidment

31762 leaves Tonbridge for Ashford, 12 October 1957. K.W. Wightman

31763 with a maroon liveried birdcage set, October 1957. K.W. Wightman

31775 passes Shortlands with a down Ramsgate express,3 August 1957.
K.W. Wightman

A pair of 'Ls', 31764 and 31766, at Tonbridge with the stock for the 2.35pm to Ashford, 24 May 1958.
K.W. Wightman

31772 at Bough Beech with a Tonbridge-Redhill local train, 31 May 1958. K.W. Wightman

31777 at Bough Beech with the 1.05pm Redhill-Tonbridge, 31 May 1958.
K.W. Wightman

31766 climbing Sole Street bank with an up Margate express, 22 August 1958.
K.W. Wightman

Chapter 4

THE 'E1' DESIGN & CONSTRUCTION

The 'Ls' had been delivered in 1914, but they failed to meet the urgent need for more powerful locomotives on the Chatham route to the Kent Coast because their weight still precluded them from several sections of the former LC&DR main line. Even the two 'E' class 4-4-0s that had been superheated were banned from that route for similar reasons. Pressure was exerted on Maunsell by the Locomotive Committee to design a locomotive with an axleload less than 18 tons that could cope with increased traffic loads over the heavily graded route, but wartime restrictions made this difficult. He did produce drawings for an outside cylindered 4-6-0 with 6ft 6in coupled wheels and an 'N' type boiler as developed in his 1915 drawings of that mogul (which again, because of the wartime production problems, did not appear until 1917). Maunsell got as far as agreeing the 4-6-0 design with the Civil Engineer subject to track alterations at Chatham and Maidstone East and the strengthening of two underbridges, but that attempt proved abortive whilst the war emergencies prevailed.

Priority was being given to the 'N' class mogul, until in 1917 the SE&CR Board decided to concentrate all the continental boat train traffic on Victoria as the route via Penge and Bickley Junction was less congested than the SER line. This decision therefore required urgent consideration of the needs of the LC&DR route. Pressure on the Civil Engineer could not produce any promise to strengthen the route to take heavier engines for at least five years after the end of hostilities. No outside manufacturer could help as they were committed to wartime munitions and supplies, so Maunsell was forced to see if anything could be done to improve existing locomotives. The heaviest engine allowed over the route was the class 'E' 4-4-0, eleven of which had been built between 1907 and 1909. Maunsell therefore investigated the possibility of developing the 'E' in conjunction with his Chief Draughtsman, James Clayton, who had been faced with similar problems in his previous role for the Midland Railway Company.

Drawings of a modified 'E' were completed in November 1917, although no progress happened until November 1918 after the Armistice when the first 'E' to be taken into Ashford Works for rebuilding was No.179. The locomotive was to be equipped with higher superheat and larger grate area and, most importantly, large diameter piston valves with ample ports and longer laps, a lesson from Churchward's Great Western engines that Maunsell had by now appreciated and adopted. Dimensions of the rebuilt engine, classified 'E1', included two new 19in x 26in inside cylinders, 10in piston valves, 6½in valve travel with 1⅝in steam lap, 6ft 6in coupled wheels, 1,504sqft heating surface (including 228sqft superheater), enlarged Belpaire firebox with 24sqft grate area, and boiler pressure of 180lbs psi. The weight of the engine was 52 tons 5 cwt, exactly the same as an un-rebuilt 'E', and the maximum axleload over the coupled wheels had actually been reduced from 17 tons 12 cwt to 17 tons 5 cwt. With a 39 ton tender, the engine's total weight was 91 tons 5 cwt. The nominal tractive effort was unchanged but they certainly were more powerful on the road with their efficient valve gear and free-steaming boiler. Five tons lighter than the class 'L', and with an axleload a full two tons lighter, the newly designated 'E1' locomotive was comparable in power with an 'L' and a lot more free-running on level track. Many of the boiler features were shared with the successful 'N' boiler.

To get the required weight loss to compensate for the additional superheat and other new features, the splasher width was reduced, cab toolboxes were removed and some heavy cast iron work replaced by fabricated metal (the toolboxes were placed on the tender). The tender capacity was 4½ tons of coal and 3,450 gallons of water. The cost of the rebuilding was just £2,711, and the livery of the new engine was a glossy black (refreshing after wartime grey). 179 was tested on a run to Headcorn in February 1919 but failed on its second outing with a major lubrication problem. After the damage had been repaired, further successful trials took place and revenue earning traffic operation began on 12 April with local working from Ashford to Hastings, before rostering to continental boat trains from 22 April. Some further alterations to improve lubrication took place in May and then teething troubles ceased and the locomotive met all expectations.

Authority was given in February 1919 by the Board for twenty more 'Es' to be converted to 'E1s' and tenders were sought from UK manufacturers. Beyer, Peacock & Co. contracted to reconstruct ten locomotives at a cost of £3,726 each, a considerably higher figure than the Ashford costing, which the government repaid in the post-war effort to stimulate Britain's manufacturing capacity. The class 'Es' selected were Nos. 19, 67, 160, 163, 165, 497, 504, 506, 507 and 511 with No.19 being rebuilt in December 1919 and the remainder between January and August 1920. The Beyer, Peacock engines were similar to the Ashford rebuilding of 179, apart from the use of Ramsbottom safety valves, and improved front-end lubrication which raised the weight slightly to 52½ tons and the axleload to 17½ tons, still within the civil engineering stipulation for the LC&DR route.

Beyer, Peacock had been asked to submit tenders for the reconstruction of twenty 'Es' and the following locomotives were selected for conversion when due for heavy repairs after November 1920 – 157, 159, 166, 176, 273, 315,

'E1' 160 rebuilt in June 1920, seen here shortly after entering traffic at Battersea (Stewarts Lane) depot, July 1920.
J.M. Bentley Collection

The first reconstruction of an 'E' in February 1919, No.179, seen here after the 1923 'Grouping' and renumbered A179, in early SR livery.
J.M. Bentley Collection

514-516 and 587. However, at the last minute it was decided to rebuild ten 'Ds' instead (as class 'D1' – see next chapter) and these class 'Es' remained un-rebuilt. In 1925, another proposal was made to rebuild all the remaining class 'Es' apart from A36 and A275 (the superheated pair), but by this time the LC&DR route had been strengthened to take 4-6-0s and the 'Scotch Arthurs' arrived on the Eastern Section for boat train work and the plan to rebuild the remaining 'Es' was aborted.

The 'E1s' were repainted dark green after the Grouping and renumbered in the SR Eastern Section 'A' series, i.e. A19, A67, etc. By 1923, sufficient track and bridge strengthening had been carried out to allow Maunsell to restore the cast iron drag boxes to improve adhesion, raising the axleload over the coupled wheels slightly to 17 tons 15 cwt. During the 1926 General Strike, four 'E1s' were converted to oil firing in June, although all were converted back to coal-burning by the end of the year. The 'E1s' were renumbered with an additional '1000' replacing the 'A' in the Southern Railway locomotive renumbering of July 1931, A19 becoming 1019 etc. Most also received new cylinders in the 1930s as their intensive use on difficult services over many years on the Chatham line was paid for in heavy wear and tear. The 'E1s' had been built with top-feed boilers but the 'D1s' had side-feed. However, as boilers were interchangeable, some 'E1s' received side-feed boilers and top-feed was phased out, the last 'E1' with top-feed being 1507 in 1948.

1163 with side-feed boiler in the post-1931 SR livery at Ashford, 31 October 1937.
MLS Collection

All the 'E1s' entered BR service at nationalisation in 1948 and all received BR numbers except for 1163, which was withdrawn in May 1949. Two more, 31160 and 31511, were withdrawn in December 1950/January 1951, but the remainder replaced older un-rebuilt 4-4-0s on semi-fast and stopping services plus summer Saturday work throughout the 1950s until 1958. 31504 and 31506 were withdrawn that year, with nearly 1.5million miles each to their credit, and 31165 was condemned in 1959. 31497 went in October 1960 and the last three, 31019, 31067 and 31507, lasted

1497 with top-feed boiler in SR livery, c1932.
F. Moore/MLS Collection

1497 after the war in somewhat rundown condition, at Stewarts Lane in company with a 'Schools', 8 September 1945. H.C. Casserley/ J.M. Bentley Collection

1506, just before nationalisation, c1947. J.M. Bentley Collection

until 1961, the final survivor being 31067 which was withdrawn in November 1961 at the same time as the last 'D1'. The lowest mileage of 1,294,124 was performed by 1163 withdrawn early, the highest were 31165 with 1,571,255 and 31179, the first convert, with 1,600,542. Fifteen out of twenty-six of the un-rebuilt 'Es' also survived until nationalisation, though most had gone by the end of 1951, with 31166 being the last to go in May 1955. 31273, the first of the un-rebuilt 'Es', managed to accumulate 1,501,994 miles, although most were in the 1.3m mile range or less.

31067 in BR mixed traffic livery in excellent condition at Stewarts Lane alongside 34070 *Manston*, 27 February 1957.
R.C. Riley/Transport Trust Collection

Operation

The first priority duty for the new 'E1' class was the haulage of continental boat trains from Victoria to Dover or Folkestone via the LC&DR route. Until that time, the most powerful locomotives allowed were the 'Es' – the 'Ls' were banned and also the two superheated 'Es'. After running in, 179 powered the 11.15am Folkestone Central-Charing Cross and the return 5.45pm Cannon Street – but only to and from Ashford. The next day, 23 April 1919, 179 was sent

The first 'E', 273, built at Ashford in February 1906, after heavy overhaul and painting in BR mixed traffic lined black livery, but before the 'lion and wheel' totem decision, c1950. It was withdrawn in October 1951.
J.M. Bentley Collection

to operate out of Battersea for the Victoria boat train services. By September all eleven 'E1s' had been delivered and were in traffic. Their initial allocation was:

Battersea: 19, 67, 165, 179, 506, 507
Dover: 160, 163
Margate: 497, 504, 511

165 was later transferred to Ashford to run comparative trials with 'D1' 727 on the proposed eighty minute scheduled Folkestone Central-Charing Cross expresses. Timings for the boat trains from Victoria to Dover Marine via Tonbridge was initially 103 minutes, later cut to 100. The maximum load on these trains was 310 tons. Some boat trains also ran via Chatham and Faversham and some – especially if running out of course – were diverted via Maidstone East.

Once accustomed to the rebuilt engines, the firemen were able to supply plenty of steam, and the engines quickly achieved an excellent reputation. There were regular boat train expresses in the early 1920s, operated by the Battersea rebuilt 4-4-0s as follows:

8.55am and 2pm Victoria-Dover Marine (Ostend ferry connections)
9.15am Victoria-Folkestone Harbour
11am and 8pm Victoria-Dover Marine
8.25am, 2.10pm and 5.45pm Dover Marine-Victoria
2.40pm and 6.40pm Dover Marine-Victoria (Ostend ferry connections)
9.10pm Folkestone Harbour-Victoria.

Battersea and Margate engines also worked expresses to Margate and Ramsgate via Faversham, including the named *Granville Express* and *Thanet Express*. They quickly ran up a substantial mileage and between mid-1920 and the end of 1922, average mileage between heavy Works repairs was around 75,000 (ranging from 60,000 to nearly 85,000) compared with the un-rebuilt 'Es' which averaged around 55,000. 506 was specially painted and kept in excellent condition for special train duties during this period. 165 was converted to oil burning in June 1921 and worked from Battersea on boat train work after initial trials, until reconversion to coal burning in March 1922.

Few records of runs in the early years exist, but Cecil J. Allen in his regular monthly articles 'British Locomotive Performance and Practice' published a couple of interesting logs in the November 1924 *Railway Magazine*. The dates of the runs were not recorded but are likely to have been around 1922/3.

Another Battersea engine, 163, on boat train work to Dover Marine, c1922. J.M. Bentley Collection

A Margate engine, 497, on a down continental boat train including a number of Pullman cars, at Wandsworth Road, c1922. H. Gordon Tidey/ J.M. Bentley Collection

Another photo of the same locomotive, 497, on a similar train at Sydenham Hill, c1922. J.M. Bentley Collection

4pm Victoria-Dover Marine
163 (E1 based at Dover)
185/195 tons

Miles	Location	Times	Speeds	Punctuality
0.0	Victoria	00.00		T
3.2	Brixton	06.25	45	½ E
4.0	Herne Hill	07.35	38*	½ E
5.7	Sydenham Hill	10.35	31/pws 40*	
8.7	Beckenham Junction	14.55	sigs 30*	T
12.6	Bickley Junction	20.40	36*	¾ L
14.9	Orpington	23.10	47½/sigs 15*	¾ E
17.7	Knockholt	30.10	sigs 15*	2¼ L
23.2	Sevenoaks	39.05	44½	4 L
28.1	Hildenborough	44.40	65	
30.6	Tonbridge	47.45	35*	4¾ L
35.9	Paddock Wood	53.15	74	4¼ L
40.5	Marden	57.10	69	
43.0	Staplehurst	59.20	74	3¼ L
46.3	Headcorn	62.05	71 ½	
51.5	Pluckley	67.30	sigs 30*/60	
57.2	Ashford	74.15	sigs 30*	4¼ L
61.5	Smeeth	78.40	66	
65.3	Westenhanger	82.15	64	
66.5	Sandling Junction	83.25	71	3½ L
70.3	Shorncliffe	86.50	71½	
71.0	Folkestone Central	87.30	55*	
72.0	Folkestone Junction	88.35	50*/68	1½ L
78.0	Dover Marine	96.00	(87 net)	1 L

O.S. Nock, drawing on records from H. Holcroft, Maunsell's assistant, reported that the 'E1s' at this time were regularly taking around 300 tons from Victoria to Dover Marine via Ashford in 99-102 minutes and developing 1,000 horse-power at 60mph, and stated that they apparently ran equally well with partly opened regulator and 25-30 per cent cut-off or full regulator and early cut-off. On a further test with 506 on a Dover-Birkenhead express between Tonbridge and Redhill, over 1,000 hp was sustained at 56mph with full regulator and 20 per cent cut-off.

After strengthening of the track bed and underbridges on the Chatham line in 1923, the superheated 'Es' and 'Ls' were permitted to run via the former LC&DR route, but because of the excellent performance of the 'E1s' and 'D1s', they retained dominance over the boat trains on both routes until the 'King Arthurs' arrived. Once their early problems had been sorted out, around 1926, the 'N15s' took over services via Tonbridge and Ashford, the Maunsell 4-4-0s retaining work on expresses to the Kent Coast via Chatham until 1932 when route restrictions between Faversham and Ramsgate were eased to permit the 'N15s' to go through to Ramsgate. Even then, with loads up to 350 tons, many men preferred the Maunsell rebuilds. The Sheerness line remained the province of the 'E1s' and 'D1s'.

Cecil J. Allen returned to the Kent Coast expresses in an article in the February 1931 *Railway Magazine* when he commented somewhat disparagingly about the lack of power available on the former LC&DR route to Margate and Dover. Although he gave no dates for any of the runs he described, I suspect they were collected from correspondents over a period of time from the mid-1920s onwards

6pm Dover Marine-Victoria
160 (E1 based at Dover)
301/315 tons (Pullman cars + baggage vans)

Miles	Location	Times	Speeds	Punctuality
0.0	Dover Marine	00.00		T
6.0	Folkestone Junction	12.20	40	¼ L
7.0	Folkestone Central	13.45		
7.7	Shorncliffe	14.50	39	
11.5	Sandling Junction	21.05	36½	T
12.7	Westenhanger	22.55		
16.5	Smeeth	-	60/55½	
20.8	Ashford	31.50	61½	¾ L
26.5	Pluckley	37.25	67	
31.7	Headcorn	42.25	69	
35.0	Staplehurst	45.30	65	½ L
37.5	Marden	47.55	64½	
42.1	Paddock Wood	52.20	62½	¼ L
47.4	Tonbridge	58.15	37*	1¼ L
49.9	Hildenborough	62.30	31½	
54.8	Sevenoaks	73.10	30*	1¾ L
	Dunton Green	75.00	60	
60.3	Knockholt	80.20	37	1¾ L
63.1	Orpington	83.35		1½ L
65.4	Bickley Junction	87.20		1¼ L
69.3	Beckenham Junction	92.55	pws 40*	2 L
72.3	Sydenham Hill	98.05		
74.0	Herne Hill	100.55		3 L
74.8	Brixton	102.20		3¼ L
78.0	Victoria	108.10 (105 net)		3 L

The specially prepared blackberry-black painted 'E1', 506, on a mid-morning Victoria-Dover Marine boat train, forerunner of the *Golden Arrow*, c1922.
John Scott-Morgan Collection

and certainly before the route was cleared for 'King Arthur' 4-6-0s and before the advent in sufficient numbers of the 'Schools'. However, trains were accelerated and the 'E1s' and 'D1s' rose to the occasion as indicated on page 56 when 'E1' 507 with an extra coach and a sharper schedule (90 minutes) was certainly speedier. A number of logs were recorded and have been made available through the archives of the Railway Performance Society. One of the most prolific recorders was a regular traveller in Kent, Mr S.A.W. Harvey, and all the remaining logs in this chapter were recorded by him unless noted otherwise.

A19 at the head of an express at Charing Cross station, c1924. John Scott-Morgan Collection

Victoria-Margate, 27.7.1930
10am Victoria-Ramsgate *(Thanet Express)*
A507 (E1-Battersea)
292/315 tons

Miles	Location	Times	Speeds	Punctuality
0.0	Victoria	00.00		T
3.2	Brixton	07.20	sigs slight	
4.0	Herne Hill	08.30		T
5.7	Sydenham Hill	11.40	sigs severe	
7.2	Penge	15.00		
8.7	Beckenham Junction	17.15		1 ¼ L
10.9	Bromley South	20.27		
12.6	Bickley Junction	23.15		1 ¾ L
14.8	St Mary Cray	26.00	65	
17.7	Swanley Junction	29.00	60	2 L
20.5	Farningham Road	31.42	75/sigs severe	
23.4	Fawkham	35.17		
26.9	Sole Street	39.15	60	2¼ L
32.9	Rochester Bridge	47.00		2 L
34.3	Chatham	49.00		1½ L
35.9	Gillingham	51.50	35*	
38.9	Rainham	55.38	60	
44.7	Sittingbourne	61.15	65	2¼ L
48.0	Teynham	64.15	67	
52.0	Faversham	68.10	69/45*	1¾ L
59.1	Whitstable	75.30	60	
62.7	Herne Bay	79.18	62/67	
70.7	Birchington	86.35	65	
72.4	Westgate	88.15	Sig stand (½ min)	
73.9	Margate	91.55	(88 net)	2 L

(Speeds not recorded but estimated from average passing times)

By the mid-1930s, the Maunsell rebuilds were used more widely. They made an occasional sortie to Brighton from New Cross Gate via the Mid-Kent line and in early 1938 1165 and 1179 were transferred to Bricklayers Arms, with 1504, 1507 and 1511 following later. 1163 and 1179 moved again later to St Leonards and began working services to Eastbourne and Brighton, although both returned to Battersea in 1939. At the beginning of the Second World War the distribution of the 'E1s' was as follows:

Battersea: 1019, 1067, 1160, 1165, 1179, 1497, 1506
Bricklayers Arms: 1163, 1504, 1507, 1511

The work of the Bricklayers Arms engines included prestigious

Victoria-Faversham, 10.7.1933
10.34pm Victoria-Canterbury East
1497 – E1
270/300 tons (Passenger and newspaper/mails)

Miles	Location	Times	Speeds	Punctuality
0.0	Victoria	00.00		T
0.7	Grosvenor Rd Bridge	02.02		
2.3	Clapham	05.30	42/40*	
3.2	Brixton	07.25	36	T
4.0	Herne Hill	09.25		½ L
0.0		00.00		
1.7	Sydenham Hill	05.25	28/30	
3.2	Penge East	08.10	48/50	
4.7	Beckenham Junction	10.00	48	1E
6.0	Shortlands Junction	12.15	40*	¾ E
6.9	Bromley South	13.55		1 E
0.0		00.00		
1.7	Bickley Junction	05.52	34	¾ L
2.5	St Mary Cray Junction	07.00	42/60	1 L
6.8	Swanley Junction	12.10	54	1 ¼ L
9.6	Farningham Road	14.50	76	
12.5	Fawkham	17.45	64	
15.0	Meopham	20.55	56	
16.0	Sole Street	22.10	50*	¼ L
20.0	Cuxton Road	26.30	56/66	½ E
22.0	Rochester Bridge	29.52	30*	¾ E
23.4	Chatham	32.20	36	¾ E
25.0	Gillingham	35.50	36	
28.0	Rainham	40.30	48	
30.7	Newington	43.52	50/sigs severe*	
33.8	Sittingbourne	48.40	sigs severe*	2¾ L
37.1	Teynham	55.00	pws 15*	
41.1	Faversham	62.30 (54.00 net)		7½ L

Stirling B1 4-4-0 No.1453 took 3 vehicles forward, 98 tons gross, from Faversham to Canterbury East.

A179 piloting a Stirling 4-4-0 on a Dover/ Ramsgate train at Victoria, c1925.
J.M. Bentley Collection

commuter runs into Surrey and Sussex suburbia. At the beginning of the war, six 'E1s' were placed in store but by December 1939 they were back in traffic and in demand for troop movement specials including a regular turn on Victoria-Poole Harbour 'Flying Boat' trains reaching the SR's Western Section at Wimbledon. Their incursion into the Western Section meant that they got used for some fill-in turns there or 'rescued' other engines in trouble, on one occasion assisting a failing 'Lord Nelson' on the 2.20pm Bournemouth West to Waterloo.

Although strictly outside the mandate for this book, I've discovered a record of a run behind the superheated 'E', 1275, during the darkest days of the war, especially in that part of England, and I reproduce it on page 62 as of interest to readers as a rare example behind one of the two superheated 'Es'

Victoria-Sittingbourne, 1.6.1935
4.35pm Victoria-Ramsgate
1179 – E1
321/346 tons

Miles	Location	Times	Speeds	Punctuality
0.0	Victoria	00.00		T
0.7	Grosvenor Rd Bridge	02.15		
2.3	Clapham	05.50	45	
3.2	Brixton	07.27	46	½ L
4.0	Herne Hill	08.40	45	¼ L
5.8	Sydenham Hill	12.02	34/30	
7.3	Penge East	14.35	50	
8.7	Beckenham Junction	16.12	62/60	¼ L
10.1	Shortlands Junction	18.00	50	
10.9	Bromley South	19.25		½ E
0.0		00.00		
1.7	Bickley Junction	06.32	24/30	1½ L
2.5	St Mary Cray Junction	07.42	45	
3.9	St Mary Cray	09.35	70	
6.8	Swanley Junction	12.46	55	1¾ L
9.6	Farningham Road	15.32	80	
12.5	Fawkham	18.18	64	
15.0	Meopham	21.26	58	
16.0	Sole Street	22.40	52*	
20.0	Cuxton Road	26.30	66	
22.0	Rochester Bridge	29.12	30*	¼ E
22.8	Rochester	30.52		T
0.0		00.00		
0.6	Chatham	03.10		1¼ L
0.0		00.00		
1.6	Gillingham	04.51		2 L
0.0		00.00		
3.0	Rainham	05.20	58	
5.7	Newington	08.22	60	
8.8	Sittingbourne	12.14		2¼ L

Described by the recorder as 'very good with this load'.

at a time when few runs were being published (I would think the recorder took some risk noting down geographic details in this part of the world at this time – he could have found himself under arrest as a spy or 'fifth columnist'!) The run was unexpectedly spirited and the 'E' matched the performance of its pre-war rebuilt versions.

Sittingbourne-Victoria, 1.6.1935
6.50pm Sittingbourne (ex-Ramsgate)-Victoria
1163 – E1
216/240 tons

Miles	Location	Times	Speeds	Punctuality
0.0	Sittingbourne	00.00		T
3.1	Newington	06.20	60	
5.8	Rainham	09.00	68	
8.8	Gillingham	12.52		¼ E
0.0		00.00		
1.6	Chatham	04.02		T
0.0		00.00		
1.4	Rochester Bridge	03.20	30*	¾ L
3.4	Cuxton Road	08.00	32	
7.4	Sole Street	15.00	40	1 E
8.4	Meopham	16.33	56	
10.9	Fawkham	19.24	75	
13.8	Farningham Road	21.50	76	
16.6	Swanley Junction	24.37	50*	2 ½ E
19.5	St Mary Cray	27.32	66	
22.3	Bickley Junction	29.55	52/60	3 E
23.4	Bromley South	32.25		3½ E
0.0		00.00		
0.8	Shortlands Junction	01.55	42	
2.3	Beckenham Hill	04.20	60	
6.0	Nunhead	09.12	46/30	¾ E
7.7	Denmark Hill	11.50	40/ pws 30*	
8.8	Brixton	14.20	32/pws 15*	¾ E
9.7	Clapham	16.35	40	
11.3	Grosvenor Rd Bridge	19.20		
12.0	Victoria	21.10	(19.30 net)	¼ L

Comfortably on time at all locations, excellent climb of Sole Street bank and good speed through Farningham Road as was customary.

1019 of Battersea on a Dover-Victoria train climbing Sole Street bank with steam to spare, 27 August 1938.
H.C. Casserley/John Scott-Morgan Collection

1160 leaves Dover Marine with a boat train for Victoria in the mid-1930s.
J.M. Bentley Collection

1511 leaves Sevenoaks Tunnel with the second portion of the 3pm Victoria-Dover (Ostend) boat train, c1936. John Scott-Morgan Collection

1019 on an up Ramsgate express at Herne Bay, c1946. H. Gordon Tidey/ J.M. Bentley Collection

1497 on a hop-pickers' special at Penshurst en route to Petts Wood, 3 September 1949. Despite it being nearly two years after nationalisation, 1497 shows no sign of belonging to British Railways. Ken Wightman

At the end of the Second World War, the 'E1s' were used on returning troop train specials and 'Demob' trains for the Midlands and North-West, via Tonbridge, Redhill and Guildford where 'N' or 'U' moguls would take over. In 1947, they were sometimes used to pilot Marsh Atlantics on Newhaven boat trains, because of temporary severe speed restrictions on that route. At the end of the war, their allocation was:

Battersea: 1019, 1067, 1160, 1165, 1179, 1497, 1511
Bricklayers Arms: 1163, 1504, 1506, 1507

1163 was withdrawn in 1949 before repainting in BR livery and three more had been scrapped by the end of 1951, but the remaining seven continued active until the late 1950s working slow and semi-fast services during the week, and relief Kent Coast expresses on summer Saturdays until they were finally withdrawn from service in 1960 and 1961. 31019 hauled an RCTS special in 1955. Four survivors (31019, 31067, 31497 and 31507) were transferred to Nine Elms in July 1959 and were sent to Salisbury to replace the ageing 'T9s', but the usual tribal conservatism was apparent, and they returned to Stewarts Lane and Bricklayers Arms in April 1960. After 31497 was condemned in October 1960,

Charing Cross-Tunbridge Wells
3.25pm Charing Cross-Hastings, 21.3.1940
1275 (superheated 'E')
190/205 tons
Bricklayers Arms Driver

Miles	Location	Times	Speeds	Punctuality
0.0	Charing Cross	00.00		T
0.8	Waterloo East	02.42		¾ L
0.0		00.00		
1.1	London Bridge	03.59		T
0.0		00.00		
3.0	New Cross	05.00	52	T
3.7	St Johns	05.50	58	
5.3	Hither Green	07.41	56/58	¼ E
8.4	Elmstead Woods	12.02	40	
9.4	Chislehurst	13.27	48	2½ E
11.9	Orpington	16.50	56	1¼ E
13.4	Chelsfield	18.42	50	
14.7	Knockholt	20.25	46	
18.7	Dunton Green	24.41	70	
20.2	Sevenoaks	26.41		3¼ E
0.0		00.00 pws 20*		
4.9	Hildenborough	08.10	70	
7.4	Tonbridge	11.31	(10 mins net)	½ L
0.0		00.00		
3.4	High Brooms	08.29	34	
4.9	Tunbridge Wells	11.30		½ L

the last three undertook local duties around London, often worked the 7.24am London Bridge-Ramsgate (the last express diagrammed for a Maunsell 4-4-0 rebuild) and 31019 ran a few Tonbridge-Brighton services in 1961.

S.A.W. Harvey was still timing trains after the Second World War, and I've selected one run in 1952 to illustrate what the 'E1s' were still capable of in their final years, the locomotive concerned being 46 years old (33 years since rebuilding).

7.08pm Sheerness-Victoria, 27.7.1952
31019 – Stewarts Lane
268/305 tons

Miles	Location	Times	Speeds	Punctuality
0.0	Sheerness	00.00		T
1.9	Queensborough	05.47		1¾ L
0.0		00.00		
2.0	Swale Halt	04.35		
4.0	Kemsley Halt	08.01		3¼ L
0.0		00.00		
1.2	Middle Junction	04.25	10* sigs	4¾ L
1.5	Western Junction	05.35	sigs*	4¾ L
3.7	Newington	12.45	35	
6.4	Rainham	16.10	64	
9.4	Gillingham	19.45	30*	
11.0	Chatham	22.45 (18 net)		8 L
0.0		00.00		
1.4	Rochester Bridge	02.53	30*	8¼ L
3.4	Cuxton Road	06.48	32 (blowing off)	
7.4	Sole Street	15.38	30	8 L
8.4	Meopham	17.16	50 (blowing off)	
10.9	Fawkham	20.11	72	
13.8	Farningham Road	23.00	70	
16.9	Swanley Junction	26.38		8 L
19.5	St Mary Cray	29.33	62	
21.7	Bickley Junction	32.20	sigs severe*	7¼ L
23.4	Bromley South	36.20	(35 net)	8¼ L
0.0		00.00		
0.8	Shortlands Junction	02.20	sigs 30 secs stand*	
2.2	Beckenham Junction	07.50		11 L
3.6	Penge East	12.02	sigs 10*	
5.1	Sydenham Hill	16.05		
6.9	Herne Hill	19.23		12½ L
0.0		00.00		
0.8	Brixton	01.54		
1.7	Clapham	03.40 (blowing off)		
3.3	Grosvenor Rd Bridge	06.40		
4.0	Victoria	08.43		12¼ L

Plenty of steam, but unable to regain time with this load after severe signal checks early and consequent loss of path after Bromley South.

31165 at St Mary Cray with a Ramsgate excursion, 17 July 1949.
Ken Wightman

Some unrebuilt 'Es' survived until this period. No.1491 working the 1.31pm Redhill-Reading train at Blackwater & Sandhurst, 12 April 1952. It was withdrawn in February 1953 without receiving its BR number or repaint.
J.M. Bentley Collection

31165 at Folkestone Warren with a Dover-London Bridge train, 1 August 1955.
L. Hanson/J.M. Bentley Collection

31497 leaving Sevenoaks Tunnel with a hop-pickers' special from London Bridge to Paddock Wood, c1953.
Ken Wightman

31019 at Victoria at the head of the RCTS *Wealden Ltd* special train, 14 August 1955. On the bufferbeam are standing famous Stewarts Lane Driver Sam Gingell and Shedmaster, Dick Hardy. Rodney Lissenden Collection

31019 on another enthusiasts' special, *The Rother Valley Ltd,* passing Kensington Olympia, c1956. John Scott-Morgan Collection

Another shot of 31019 on the *Rother Valley Ltd* passing East Acton, c1956. J.M. Bentley Collection

31067 on a Hastings-Ashford local nearing its destination, c1958.
J.M. Bentley Collection

Personal Reminiscences of the 'E1s'

Unfortunately, this section of the book is exceedingly short. I never had a run behind an 'E1', didn't even experience a 'near miss'. My trusty *ABC of Southern Locomotives* (1/6d), July 1947, which had sold 180,000 copies to trainspotters since the first edition in December 1942, identified just one 'E1' that I'd underlined, 1067. Where had I seen that? I presume on our 1946 summer holiday to Brighton on the *Brighton Belle,* when I first started trainspotting at the instigation of a father trying to occupy a bored eight-year-old impatient to get to the seaside. The following year we went to Margate, which sounds promising, but unfortunately, we went by coach because we couldn't afford the train fare. Perhaps I was allowed to venture to Margate station, but I have no recollection of that.

By 1951, I see that I'd marked off 31019, 31067, 31504 and 31506, as well as 'Es' 31166 and 31315, the last two survivors of the un-rebuilt engines. I don't remember ever going to Victoria, London Bridge or Charing Cross in those early years, but I did stop off a couple of times at Clapham Junction and perhaps I saw them there or had a brief glimpse down into the depths of Stewarts Lane as my local electric from Hampton Court passed Queen's Road and Nine Elms en route to a day round the London termini. We were always glued to the window to catch a fleeting sight of anything standing near enough to grab its number.

I never did see 31507 while I was still a 'locospotter', but in 1959 the University College London Railway Society had shed visits of Stewarts Lane and Bricklayers Arms and I managed a (not very good) photo of 31507 on Bricklayers Arms shed, along with a number of 'Schools' and one of the first rebuilt 'West Countries'. 31507 lasted another two years until withdrawal in July 1961, being condemned just four months before the last survivor, 31067. Dick Riley and his friends, Ken Wightman and David Clark, were busy taking colour photos at this time, especially in the Shortlands Junction, Beckenham area where they lived, and the only two colour slides I can trace from them in those latter years were both of 31507, following on page 67.

'E1' 31507 in its last years, 1958-60, then leaves Wadhurst with a London Bridge-Hastings train, c1958. Ken Wightman

At London Bridge on the 12.44pm parcels train to Ramsgate, c1960. R.C. Riley Collection

Chapter 5

THE 'D1' DESIGN & CONSTRUCTION

Maunsell had gained authority to rebuild ten more 'Es' in 1919 (as recounted in Chapter 4) but Beyer, Peacock could only promise to complete the original order of ten. However, by 1921 the company advised that they were now in a position to meet the outstanding order. Authority to contract for this was agreed in 1920, but before it was put into effect, Maunsell was advised by the Works management that the same boiler and cylinders could be fitted to the 'Ds' without further modification and he gave the order to rebuild the following 'Ds': 246, 247 (the 'D' with the extended smokebox), 487, 489, 494, 502, 545, 735, 747 and 749.

Dimensions of the new 'D1s' were similar to the 'E1s' except for the following: 6ft 8in coupled wheels compared with 6ft 6in; 3ft 7in diameter bogie wheels instead of 3ft 6in; sight-feed instead of mechanical lubrication; total weight of only 51 tons 5 cwt; and axleload of 16¾ tons. The ten engines were rebuilt during 1921 and were delivered and received into traffic between April and November of that year, 246 and 247 being first, 489 and 749 being the last. Ashford rebuilt two more, 145 and 727, in 1922.

At the Grouping, all were renumbered with an initial 'A' as were all South Eastern Section locomotives, and after the strengthening of the LC&DR route in 1923, their adhesion was improved by fitting heavy cast-iron drag boxes below the footplate, increasing the axleload to 17 tons and the engine weight to 52 tons 4 cwt. Boilers became interchangeable with the 'E1s' and thus a mixture of top and

494, rebuilt by Beyer, Peacock & Co. in July 1921, seen here shortly afterwards, c1921. John Scott-Morgan Collection

side-feed of water to the boiler. Mileages worked between heavy repairs (around 71,500) were similar to the 'E1s' although their older frames needed more repair work as they aged.

In 1925 it was decided to rebuild all the remaining 'Ds' at a cost of £2,790 each, but because the remaining parts of the LC&DR route had now been strengthened to permit the passage of the 'King Arthur' 4-6-0s, the rebuilding was thought to be unnecessary and too costly and only nine had been rebuilt before the order was rescinded. The locomotives that were completed at Ashford between

735, also rebuilt by Beyer, Peacock & Co. in July 1921, a year later, c1922. F. Moore/ J.M. Bentley Collection

A247, rebuilt from the unique 'D' with extended smokebox by Beyer, Peacock & Co in March 1921, photographed at Ashford, 25 August 1927. It was condemned in July 1961. H.C. Casserley/J.M. Bentley Collection

A470, rebuilt at Ashford in November 1926 as part of the second batch of reconstructions, photographed at Battersea Queens Road, c1927. J.M. Bentley Collection

November 1926 and July 1927 were: A470, A492, A505, A509, A736, A739, A741, A743 and A745. Six of these (492, 505, 509, 739, 741 and 743) were given new frames and were thus virtually new engines.

After the 1931 SR renumbering, 1,000 was added to their numbers in place of the prefix 'A'. Frames of some of the earlier rebuilds were now beginning to show heavy wear, requiring large patches, and three new sets of frames were built at Ashford, one-sixteenth inch thicker than the original 'D' frames and the first was allotted to 1502 in 1938. The other two were held in reserve and were used for part replacement on three further 'D1s' – 1246, 1487 and 1545.

All the original un-rebuilt 'Ds' had been withdrawn by 1956.

1727 at Bricklayers Arms, in company with another 'D1' and 'N15', c1946. J.M. Bentley Collection

Six managed to hang on to that year, mostly operating three coach stopping services on the Reading-Guildford-Redhill line – 31075, 31488, 31549, 31574, 31577 and 31737. The last two, withdrawn in December, were 31075 and 31577. 31737, withdrawn in November, was selected for preservation and restoration and is one of the National Railway Museum's collection at York.

Most of the rebuilt 'D1s' outlasted them, although 1747 was scrapped in 1944 after sustaining war damage. The rest were received into BR ownership after nationalisation and were renumbered with the addition of 30,000 as for most SR locomotives. The SR livery gave way to BR mixed traffic lined black, 31749 being repainted in December 1948, the last to receive this livery being 31246 in August 1953. 31502 and 31745 never received the lined livery, being plain black when condemned. 31736 was the first to go, apart from 1747, in December 1950 and 31502 and 31745 were withdrawn in 1951. However, the others continued to receive heavy overhauls despite the availability of an increasing number of Bulleid pacifics and Brighton-built LMS 2-6-4Ts. The completion of electrification of the Kent Coast line via Chatham in June 1959 led to their redundancy and 31470 and 31741 were withdrawn. Dispersal of the others elsewhere, though often into store, prolonged their lives for a while, but four more, 31492, 31494, 31509 and 31743, were condemned in 1960 and the rest were withdrawn in 1961, the last survivors being 31489, 31739 and 31749 in November.

1487 at Dover, 1935. Photomatic/J.M. Bentley Collection

s1735 at Dover with BR lettering, black livery, the 's' denoting BR Southern Region before the 30,000 was added, 14 July 1950. H.C. Casserley/J.M. Bentley Collection

31741 at Bricklayers Arms, 29 March 1953. MLS Collection

31494 taking water, c1957. J.M. Bentley Collection

31727 at Redhill, in BR mixed traffic livery with the large BR icon on the tender, 18 February 1958. MLS Collection

31735 stored awaiting a withdrawal decision at Eastleigh, 4 March 1961. It was condemned the following month. J.M. Bentley Collection

31145 at Stewarts Lane, 2 August 1959. R.C. Riley

31545 at Stewarts Lane, with a Maunsell 'W' 2-6-4T behind, 10 May 1959. R.C. Riley

31743 at Ramsgate, 28 March 1959. R.C. Riley

A view from above of 31489 on Ramsgate shed, 14 May 1960. R.C. Riley

A side view of 31749 at Stewarts Lane, 30 March 1959. R.C. Riley

Operation

After running in, the initial allocation of the 1921/2 rebuilds was as follows:

Battersea: 247, 502, 747
Dover: 145, 487, 489
Bricklayers Arms: 727, 749
Margate: 246, 494, 545, 735

They appear to have worked turn and turn about with the 'E1s' with little difference in performance. 545 began to head the 8.05am Margate-Cannon Street and 6.12pm return which had been previously the responsibility of double-headed 'Ds'. Other regular turns involved the *Thanet Pullman Ltd*, a train weighing 238 tons, running the 73 miles non-stop between Victoria and Margate at an average of 50mph in both directions, despite the many speed restrictions on the route.

In 1922, the SE&CR conducted a series of trials on the 8.39am Ashford to Cannon Street and the 3pm Charing Cross-Ashford, with an 'L', the lone 'K' 2-6-4T No. 790, and a 'D1', Maunsell's assistant, H. Holcroft, travelling on all three locomotives. 'D1' 749 was tried for a week and outperformed the 'L' especially on the Tonbridge-Ashford section. The 'L' and the 'K' were comparable, the 'L' having the superior boiler and the 'K' having the more efficient cylinder/valve system, but the 'D1' was the most free-running and competent machine suitable for the SER's longer distance work. More 'River' class 'K' 2-6-4Ts were built between 1925 and 1927, but they were heavy and never displaced the 'E1s' and 'D1s' which held sway on the LC&DR route until the 'King Arthurs' eventually replaced them on the heaviest and most prestigious services.

Three new cross-country services were introduced in 1922, Ramsgate-Manchester/Liverpool, Deal/Margate-Birkenhead and Dover-Bournemouth West. The 'D1s' were rostered to these services, to Kensington and Redhill and although at first an 'N' 2-6-0 took over the Birkenhead train after reversal at Redhill until it was deemed too rough riding at the speeds required and a 'D1' worked that forward section also. The Bournemouth service was combined with the Birkenhead train as far as Guildford where a LS&WR 'T9' took over running via Portsmouth and Southampton.

From the late 1920s through the 1930s, with 'King Arthurs' (and a few 'Lord Nelsons') on the boat trains, the rebuilt 4-4-0s maintained their work on the Sheerness and Kent Coast trains from Charing Cross, Cannon Street and Victoria, until the 'Schools' arrived. The first ones replaced the 'Ls' on the faster Hastings expresses and shared Kent Coast/boat train work with the 'King Arthurs', the 'D1s' and 'E1s' retaining semi-fast and stopping train work as well as the cross-country trains previously mentioned. In 1937, the allocation of 'D1s' was:

Dover: 1145, 1246, 1470, 1487, 1545, 1727, 1749
Ramsgate: 1735, 1736, 1741, 1743, 1745
Faversham: 1247, 1489, 1492, 1494, 1502, 1505, 1509, 1739, 1747

When the *Night Ferry* (Dover-Dunkirk) began to run in 1936, it was usually hauled by a pair of 4-4-0s of any of the 'Ls', 'D1s', 'E1s' or 'L1s' and later after the Second World War, one of the 4-4-0s would normally pilot a Bulleid light pacific. Some of the 4-4-0s were stored during the winter in the late 1930s as the Maunsell 'Schools' were produced in more numbers and took over key expresses to Hastings

A487 of Dover on the all Pullman *Thanet Pullman Limited*, c1923. John Scott-Morgan Collection

A247, a Battersea engine, on an express for Ramsgate and Dover, c1925. F.E. Mackay/ J.M. Bentley Collection

Wainwright 'D' 728 pilots an unidentified 'D1' on a heavy down continental boat train approaching Tonbridge station, c1923. The Redhill branch is to the left and the layout is before through middle roads were laid in 1925. J.M. Bentley Collection

and the Kent Coast, but the 4-4-0s were still required for semi-fast and stopping services and in summer for relief Kent Coast expresses. In May 1938, three 'D1s' (1492, 1494 and 1502) were transferred from Faversham to Battersea and began to run on some non-electrified Central Section services allocated to the former SE&CR shed.

In the February 1931 article previously mentioned (see page 54) C.J.Allen also included runs with 'D1s' as well as 'E1s'. He described four runs on non-stop Victoria-Margate expresses, probably the 10am Victoria *Thanet Pullman,* allowed 97 minutes for the 73.9 miles, with a load of seven coaches plus Pullman car, weighing 267 tons (285 tons gross). The net time ranged between 94 and 96 minutes for three of them, with no higher speed than the upper 60s at Farningham Road and just reaching 70mph at locations like Sittingbourne and descending from Herne Bay to Birchington. They were able to meet the schedule but with little in hand for out of course delays. One run with A509 was distinctly better, achieving the run in 94 minutes 25 seconds (2½ minutes early) – 89¾ minutes net – with a top speed of 72mph at Farningham Road. Cecil J. Allen also published a couple of up runs, one with A246 which just held the 97 minutes schedule, but averaged 67mph from Fawkham to Farningham Road and probably reached 75mph at the latter point. Both the 'D1s' (the other was A505) fell to the mid-20s on the ascent of Sole Street bank.

Also, a number of runs recorded by S.A.W. Harvey and retrieved from the Railway Performance Society's archives, from which I've selected a number of typical runs and a couple of excellent performances. I'll start with one such very competent performance on the evening Cannon Street-Margate express in 1933.

Cannon Street-Margate
5.04pm Cannon Street, 4.7.1933
1741 – Ramsgate
252/275 tons

Miles	Location	Times	Speeds	Punctuality
0.0	Cannon Street	00.00		T
0.7	London Bridge	02.25	30	T
3.7	New Cross	06.50	46	¾ E
4.4	St Johns	08.15	pws 40*	¼ E
5.0	Parks Bridge Jcn	09.25	pws 40*	T
6.0	Hither Green	10.50	44	¼ E
7.8	Grove Park	13.45	40	
9.1	Elmstead Woods	16.05	38	
10.1	Chislehurst	17.50	sigs 30*	1 ¼ E
12.4	St Mary Cray	22.25	38/ sigs 36*	
15.3	Swanley Junction	26.20	52	¾ E
18.1	Farningham Road	29.07	68/80	
21.0	Fawkham	31.40	72/66	
23.5	Meopham	34.30	60 easy	
24.5	Sole Street	35.40	52/70	1¼ E
28.5	Cuxton Road	39.50	54*	2¼ E
30.5	Rochester Bridge Jcn	42.58		2½ E
31.3	Rochester	44.25	38*	
31.9	Chatham	45.24	40	2½ E
33.5	Gillingham	50.30	sigs 20*/35	
36.5	Rainham	54.35	64	
39.2	Newington	57.15	66	
42.3	Sittingbourne	60.08	76	1 E
45.6	Teynham	62.58	74/70	
49.6	Faversham	67.10	40*	1¾ E
56.7	Whitstable	75.00	70	
58.1	Chestfield	76.42	60	
60.3	Herne Bay	79.00	62/78	2 E
68.3	Birchington	87.00	74	
70.0	Westgate	88.48	64	
71.5	Margate	91.30	(86.45 net)	1½ E

Next is an up run from Folkestone to Charing Cross, again timed by S.A.W. Harvey, again competent, but this time with a lighter load of only 160 tons gross. Although engine renumbering had taken place in 1931, the 'D1' at the head of this train was still numbered A246.

Folkestone Central-Charing Cross
5.54pm Folkestone, 31.7.1932
A246 – Dover
149/160 tons

Miles	Location	Times	Speeds	Punctuality
0.0	Folkestone Central	00.00		T
0.7	Shorncliffe	02.20	32	
1.2	Cheriton	03.20	34/44	
4.5	Sandling Junction	07.45	50	¼ E
5.7	Westenhanger	09.12	54	
9.5	Smeeth	12.57	70	
13.8	Ashford	16.50	76	¼ E
19.5	Pluckley	22.15	70/65	
24.7	Headcorn	26.57	72	
28.0	Staplehurst	29.50	74/69	
30.5	Marden	32.10	74	
35.1	Paddock Wood	36.10	72	T
40.4	Tonbridge	41.28	32*	½ E
42.9	Hildenborough	45.22	44/ sigs 24*	
47.8	Sevenoaks	54.23	38/50	½ E
49.3	Dunton Green	56.05	65	
53.3	Knockholt	61.35	40	½ E
54.6	Chelsfield	63.05	64	
56.1	Orpington	64.35	66	½ E
57.3	Petts Wood	65.50	60	
58.6	Chislehurst	67.22	56	½ E
59.6	Elmstead Woods	68.27	58	
60.9	Grove Park	69.45	66	
62.7	Hither Green	71.27	68	½ E
64.3	St Johns	73.20	50	
65.0	New Cross	74.55	sigs 36*	T
68.0	London Bridge	79.35		½ E
69.1	Waterloo East	83.20	36	
69.9	Charing Cross	85.15	(81.15 net)	¾ E

Many of the logs recorded on both the former SER and LC&DR routes seemed to be plagued with frequent signal checks indicating the congested nature of both routes. However, despite loss of time because of the checks, many of S.A.W. Harvey's logs are annotated 'blowing off steam' in the margin and it seems apparent that Maunsell's 'E1s' and 'D1s' were very free-steaming machines. In the run below, the sudden burst of speed after Sevenoaks appears to have been a sign of frustration after so many checks, and the momentary 84mph is the highest speed I have discovered by the rebuilt SE&CR 4-4-0 from the logs I have perused.

Charing Cross-Paddock Wood
1.15pm Charing Cross, 17.3.1934
1487 – Dover
256/270 tons

Miles	Location	Times	Speeds	Punctuality
0.0	Charing Cross	00.00		T
0.8	Waterloo East	03.10	sigs*	1¼ L
0.0		00.00		
1.1	London Bridge	04.10	sig stand 15 secs	¼ L
4.1	New Cross	10.25	sigs 15*	1½ L
4.8	St Johns	11.55	38	
6.4	Hither Green	14.48	46	
9.5	Elmstead Woods	20.20	40	
10.5	Chislehurst	22.00	46	4 L
13.0	Orpington	25.52	52	4¾ L
14.5	Chelsfield	28.30	sigs 10*	
15.8	Knockholt	33.50	30	
19.8	Dunton Green	39.12	74/70	
21.3	Sevenoaks	40.43	60	9¾ L
26.2	Hildenborough	45.25	72/84	
28.7	Tonbridge	48.15	46*/sigs	9¼ L
34.0	Paddock Wood	55.00	(44.45 net)	9 L

I now record three runs behind 'D1s' with similar heavy loads for the former LC & DR route.

Victoria-Chatham

		2.04pm Victoria, 5.5.1934 1545 – Dover 297/322 tons			**4.34pm Victoria, 2.10.1937 1492 – Faversham 306/320 tons**			**4.34pm Victoria, 29.4.1939 1505-Faversham 256/270 tons**		
Miles	**Location**	**Times**	**Speeds**		**Times**	**Speeds**		**Times**	**Speeds**	
0.0	Victoria	00.00		T	00.00		T	00.00		T
0.7	Grosvenor Rd Bridge	02.09			02.13			02.23		
2.3	Clapham	05.37	48/42		06.05	46/40		06.10	45/40	
3.2	Brixton	07.20	48	¼ E	07.58	44		07.48	45	¼ L
4.0	Herne Hill	08.32	46	½ E	09.20	sigs 20*	¼ L	09.05	40	T
5.7	Sydenham Hill	12.15	sigs 30*		13.37	22		12.27	35/32	
7.2	Penge East	15.10	50		16.50	45		14.58	48	
8.7	Beckenham Junction	16.52	64	1 ¼ E	18.42	56	1¾ L	16.50	sigs sl*	¼ E
10.1	Shortlands Junction	18.50	sigs*	¾ E	20.40	50		18.48	50	¼ L
10.9	Bromley South	20.25		1 ½ E	22.06	(21 net)	1 L	20.16		¾ E
0.0		00.00			00.00			00.00		
1.7	Bickley Junction	05.59	40	1 L	07.30	34	2 ½ L	05.18	35	¼ L
3.9	St Mary Cray	08.49	50/72		10.37	48/68		08.19	45/65	
6.8	Swanley Junction	12.00	56	1 L	14.04	50	3 L	11.50	pws *	¾ L
9.6	Farningham Road	14.44	82		16.57	78		15.27	72	
12.5	Fawkham	19.00	pws 40*		19.45	60		18.30	56	
15.0	Meopham	23.25	60		22.55	52		21.55	52	
16.0	Sole Street	24.24	56	3¾ L	24.10	50	3¼ L	23.13		2¼ L
20.0	Cuxton Road	28.46	68	2¾ L	28.07	70	2 L	27.10	78	1¼ L
22.0	Rochester Bridge	31.55	sig stands		30.35	35*	1 L	29.35	35*	T
23.4	Chatham	41.28	(31 net)	9½ L	32.08		1 L	31.08	(30 net)	T

Ramsgate's 1743 at Kensington Addison Road with a through train from the LMS for Redhill and Dover, 26 August 1933. H.C. Casserley/ J.M. Bentley Collection

London Bridge-Ashford
7.24am London Bridge-Ramsgate, 17.8.1935
1741 – Ramsgate
166/180 tons

Miles	Location	Times	Speeds	Punctuality
0.0	London Bridge	00.00		T
3.0	New Cross	04.44	54	
3.7	St Johns	05.35	54	
5.3	Hither Green	07.40	44	1¼ E
8.4	Elmstead Woods	12.45	40	
9.4	Chislehurst	14.40	sigs 15*	2¼ E
10.7	Petts Wood	17.37	45	
11.9	Orpington	19.45 (18.30 net)		2¼ E
0.0		00.00		
1.5	Chelsfield	03.25	35	
2.8	Knockholt	05.30	40	
6.8	Dunton Green	09.58	64/70	
8.3	Sevenoaks	11.49		1¼ E
0.0		00.00		
4.9	Hildenborough	06.27	64/80	
		-	sig stand 1m 55s	
7.4	Tonbridge	13.20	(9m 45s net)	2¼ L
0.0		00.00		
5.3	Paddock Wood	07.15	68	
9.9	Marden	11.34	60	
12.4	Staplehurst	14.00	64	
15.7	Headcorn	17.10	62	
20.9	Pluckley	22.28	60	
26.6	Ashford	29.20		1½ L

This train was a regular 4-4-0 turn right up to the end of the classes in 1961. It was booked for an engine change at Ashford and 1741 was replaced by an 'L1', No.1782. My last pre-war record (again by S.A.W. Harvey) is of a 'D1' on one of the trains that ran to Redhill and then diverted to the route to Reading, Oxford and the West Midlands. From Folkestone to Ashford the load was only three coaches for 96 tons tare and time was easily kept. The train was made up at Ashford and the log commences there.

1489 with a Kent Coast express at Bromley South, 15 August 1936.
H.C. Casserley/J.M. Bentley Collection

Dover's 1487 pilots an 'L1', No.1756, on the up *Night Ferry* at Shorncliffe, 1936. A 'Nord' *fourgon* is in the consist.
J.M. Bentley Collection

Ashford-Redhill
10.01am Folkestone, 15.6.1938
1487 – Dover
193/205 tons

Miles	Location	Times	Speeds	Punctuality
0.0	Ashford	00.00		T
5.7	Pluckley	08.26	65	
10.9	Headcorn	13.27	65	
14.2	Staplehurst	16.30	65	
16.7	Marden	19.00	58	
21.3	Paddock Wood	23.25	64	½ L
26.6	Tonbridge	29.43		¼ E
0.0		00.00		
2.5	Lyghe Halt	05.04		
4.2	Penshurst	07.24	54	
9.2	Edenbridge	13.17	54	
11.5	Crowhurst	16.05	54	
14.1	Godstone	19.05	50	
17.6	Nutfield	23.15	52	
19.7	Redhill	27.13		¾ E

These trains were more easily timed than the trains on the main London routes and 1487 was able to maintain punctuality without overmuch effort.

At the commencement of the Second World War, eight 'D1s' were put to store, but were restored to traffic for Christmas parcels and mails and the Dunkirk evacuation. Then, when the Southern was required to augment the freight haulage capacity of other railways by supplying 'S11s', 'S15s', 'H15s' and 'N15Xs', five 'D1s' (1145, 1247, 1492, 1494 and 1739) replaced some of them at Nine Elms and were working Waterloo-Basingstoke-Salisbury semi-fast trains and parcels and empty stock trains on the Bournemouth route. Two were

Faversham's 1502 at Brixton on a semi-fast train from Sheerness composed of SE&CR six-wheel rolling stock, c1938. J.M. Bentley Collection

1502 arrives at Charing Cross with an express from the Kent Coast, 30 May 1938.
H.C. Casserley/J.M. Bentley Collection

Ramsgate's 1736 passing Bromley South with the two-coach winter portion of the *Sunny South Express*, 22 October 1938.
H.C. Casserley/J.M. Bentley Collection

loaned to Guildford (1145 and 1492) for a Woking-Southampton Terminus service and they were involved in troop train haulage. There were also Victoria-Poole Harbour boat trains for the 'Flying Boat' air services, which they shared with the 'E1s'.

After the war, and the return of the engines loaned to the Southern's Western Section, the allocation at the end of 1945 was:

Battersea: 1145, 1247, 1492, 1494, 1735, 1736, 1743, 1745, 1749
Dover: 1545, 1727
Faversham: 1246, 1470, 1487, 1489, 1502, 1505, 1509, 1739, 1741

1747 of Faversham depot at Bromley South with a down Kent Coast express, 18 April 1938.
H.C. Casserley/J.M. Bentley Collection

Many Battersea and Dover 'D1s' worked military specials at the end of the war as the army was disbanded – especially from Folkestone Harbour back to the West Midlands and North via Redhill, where usually Maunsell 2-6-0s would take over for the run to Reading via Guildford or Kensington.

It is interesting to see how the 'D1s' performed in the immediate post-war period when maintenance had been neglected. After nationalisation, the 4-4-0s were mainly employed on slow stopping trains which did not attract the train timers – the only opportunities of interest were mainly on summer Saturday holiday trains when lines were congested and signal checks frequent. However, O.S. Nock timed the 5.22pm Cannon Street-Rochester 8-coach 275 ton gross train in 1948 with 1509 and made the 31.3 miles in 41 minutes 46 seconds (schedule 45 minutes) including a signal check at Grove Park. Unusually high speeds were reached for the immediate post-war period – 76mph at Farningham Road and 78mph at the foot of Sole Street bank.

S.A.W. Harvey recorded very similar net times on the same train, basically a commuter service, and again recorded speeds in the mid to high 70s.

1747, the 'D1' destroyed by bombing in the Second World War, at Elephant & Castle with an up Kent Coast express, c1939. J.M. Bentley Collection

Cannon Street-Chatham

		5.22pm Cannon Street, 14.8.1947 1545 – Dover 254/274 tons			5.22pm Cannon Street, 30.8.1948 31487 - Faversham 247/270 tons		
Miles	**Location**	**Times**	**Speeds**	**Punctuality**	**Times**	**Speeds**	**Punctuality**
0.0	Cannon Street	00.00		T	00.00		T
0.7	London Bridge	02.46	sigs 10*	¼ L	02.14	sigs	¼ E
3.7	New Cross	08.30	sigs	1½ L	07.12	sigs	¼ L
4.4	St Johns	09.30	blowing off	1 L	08.15		¼ E
6.0	Hither Green	11.45		1¼ L	10.45		¼ L
7.8	Grove Park	14.28	40		13.40	40	
9.1	Elmstead Woods	16.37	38		15.49	40	
10.1	Chislehurst	18.05	44	1½ L	17.20	35	¾ L
11.1	St Mary Cray Jcn	19.32	45	1 L	19.15	60	¾ L
15.1	Swanley	26.00	pws 15*	2½ L	24.20	54	¾ L
18.2	Farningham Road	29.30	74		27.22	75	
21.1	Fawkham	32.25	48		29.59	62	
23.6	Meopham	35.39	50		32.58	52	
24.6	Sole Street	36.51	50	1¾ L	34.09	55	¾ E
28.6	Cuxton Road	40.29	78	½ L	38.15	62	1¾ E
30.6	Rochester Bridge	43.50	30*	¼ E	47.10	sig stand 4 ½ mins stand)	
31.4	Rochester	46.14	sigs 5* (41 net)	¾ L	49.50	(41 net)	4½ L
0.0		00.00			00.00		
0.6	Chatham	02.54		1½ L	03.20		5¾ L

31246 at Cheriton with an up Kent Coast express, 4 August 1952. L. Hanson/J.M. Bentley Collection

31246 on a Dover-Ashford stopping train at Folkestone Warren, 29 July 1955. L. Hanson/J.M. Bentley Collection

An unidentified 'D1' assists a 'Battle of Britain' which is slipping badly at Shortlands Junction on the up *Night Ferry*, c1956. Ken Wightman

31509 leaving Lyghe with a Redhill-Tonbridge train, c1957. Ken Wightman

Chatham-Victoria
9.25am Gillingham, 16.1.1948
1487 – Faversham
161/170 tons

Miles	Location	Times	Speeds	Punctuality
0.0	Chatham	00.00		T
0.6	Rochester	02.02		T
0.0		00.00		
0.8	Rochester Bridge	02.05		
2.8	Cuxton Road	06.35		
6.8	Sole Street	14.45	34	¾ L
7.8	Meopham	16.15		
10.3	Fawkham	19.02	70	
13.2	Farningham Road	22.15	pws 20*	
16.3	Swanley	28.30		2½ L
18.9	St Mary Cray	31.42		
21.1	Bickley Junction	34.35		2½ L
22.8	Bromley South	37.10	(34½ mins net)	1¼ L
0.0		00.00		T
0.8	Shortlands Junction	01.43		
2.2	Beckenham Junction	03.48		2½ E
3.6	Penge East	05.52		
5.1	Sydenham Hill	08.40		
6.9	Herne Hill	11.10	sigs/SL	2¾ E
7.7	Brixton	12.53		2¼ E
8.6	Clapham	14.20		
10.2	Grosvenor Bridge	16.46		
10.9	Victoria	18.45	(18 mins net)	2¼ E

Three 'D1s' were withdrawn in 1950/1, after achieving mileages in the 1.3million region, but the remainder received Works overhauls and – despite the large number of Bulleid light pacifics that took over many turns on both Kent Coast routes – found employment on summer Saturdays on relief holiday trains, and were occupied on weekdays on secondary duties with 'D1s' transferred to Tonbridge (31246, 31509, 31727 and 31735) and Gillingham (31492, 31494, 31745) sheds, although some inroads in

		Victoria-Herne Bay, 13.9.1952 1.50pm Victoria – Margate 31487 – Faversham 256/275 tons			**Victoria-Newington, 27.7.1952 10.50am Victoria-Sheerness 31505 - Faversham 268/290 tons**		
Miles	**Location**	**Times**	**Speeds**	**Punctuality**	**Times**	**Speeds**	**Punctuality**
0.0	Victoria	00.00			00.00		T
0.7	Grosvenor Rd Bridge	02.15			02.02		
2.3	Clapham	05.40			05.30		
3.2	Brixton	07.43	sigs 10*	¾ L	07.15	sigs 15*	¼ L
4.3	Denmark Hill	09.47	50		09.05	50	
6.0	Nunhead	12.15	45	¼ L	11.21	54/48	¾ E
8.1	Catford	15.20	50 (blowing off)		14.31		
9.7	Beckenham Hill	17.19	50/46		16.50		
11.2	Shortlands Junction	19.30	40*	T	19.59	30*	T
12.0	Bromley South	20.35			21.17	sigs	
13.7	Bickley Junction	23.42		1¼ E	25.05		1 E
14.5	St Mary Cray Junction	24.45	65	1¼ E	26.15	60	
18.5	Swanley	29.39	50	1¼ E	31.40	45	2½ E
21.6	Farningham Road	34.15	pws 15*		35.39	70	
24.5	Fawkham	38.25	45		39.02	52	
27.0	Meopham	42.13	48		43.10	45	
28.0	Sole Street	43.35		½ E	44.40		2¼ E
32.0	Cuxton Road	48.11	50/56	1¾ E	50.15	50*	1¾ E
34.0	Rochester Bridge	53.35	sig stand	1 E	53.35		2½ E
34.8	Rochester	56.00			54.55		
35.4	Chatham	57.40		¾ L	56.07	30*	3 E
37.0	Gillingham	63.20	sigs 5*		59.44	30*	
40.0	Rainham	68.08	58		64.30	50	
42.7	Newington	71.15	65		67.45	50 (blowing off)	
43.8	MP 42 ¾	72.37	51			To Sheerness	
45.8	Sittingbourne	75.05		4 L			
49.1	Teynham	78.47	55				
53.1	Faversham	84.52	48/30*	5¾ L			
56.2	Graveney	88.40	58				
60.2	Whitstable	92.45	62				
61.6	Chestfield	94.30	55/sigs				
63.8	Herne Bay	98.08	(86 mins net)	4 L			

this type of work were made by the Brighton-built LMS 2-6-4Ts.

31545 worked a Stephenson Locomotive Society Special in May 1957 and its driver was the legendary Sammy Gingell of Stewarts Lane shed.

By 1956, the end was coming for the un-rebuilt 'Ds' with the remaining three (31075, 31488, 31577) undertaking stopping services mostly on the Redhill-Reading line, concentrated there after earlier local work around Faversham had become the domain of the rebuilt 4-4-0s.

A couple of runs on the 7.12pm Margate-Victoria were recorded in the summer of 1957

Victoria-Margate, 19.5.1957
10.58am Victoria SLS Special
31545 – Stewarts Lane
6 chs 192/210 tons
Driver S.Gingell, Fireman J.Williams

Miles	Location	Times	Speeds	Punctuality
0.0	Victoria	00.00		¼ L
0.7	Grosvenor Rd Bridge	-	17	
	Factory Junction	04.39	30/40	
3.2	Brixton	06.54	36½	¼ L
4.0	Herne Hill	08.16	33/31	T
5.8	Sydenham Hill	11.30	41	
7.3	Penge East	13.49	50	
8.7	Beckenham Junction	15.47	40½	½ L
10.1	Shortlands Junction	17.37	55	
10.9	Bromley South	18.39	42/36	
12.6	Bickley Junction	21.13	46/53	T
14.8	St Mary Cray	23.52	58	
17.7	Swanley	26.42	60/73	T
20.5	Farningham Road	29.33	77/64	
23.4	Fawkham	32.04	55/61	
25.9	Meopham	34.37	60	
26.9	Sole Street	35.38	73/pws 10*	¼ E
30.9	Cuxton Road	41.17	53/27*	1 L
32.9	Rochester Bridge Jcn	44.41	32	1 L
34.3	Chatham	47.21	37	1 L
35.9	Gillingham	49.56	48/55	
38.9	Rainham	53.37	61	
	Newington	56.18	52/68	
44.7	Sittingbourne	59.24	75	1¼ L
48.0	Teynham	62.02	71/60 eased	
52.0	Faversham	66.01	34*/57	¾ L
	Graveney Sidings	69.51	61/55	
59.1	Whitstable	72.08	50	
	Chestfield	75.55	64	
62.7	Herne Bay	78.03	49½ / 72	¼ L
	Reculver Box	81.29	79	
70.7	Birchington	85.03	67	
72.4	Westgate	86.28	72	
73.9	Margate	88.04	(85.30 net)	1¾ E

with overloaded 'D1s' returning holiday makers to London.
On the first, on 7 July, 31545, with 320 tons gross, struggled early on despite showing evidence of having plenty of steam (the recorder noted that it was blowing off steam on several occasions throughout the journey). It dropped four minutes on the eleven mile section to the Herne Bay stop, falling to 30mph on the 1 in 96 to MP 64¼. It then made a reasonable fist of it, holding schedule to Faversham without exceeding 50mph and dropping just one minute on to Sittingbourne. A good ascent was made to the summit of the 1 in 95/120 after Sittingbourne averaging 43mph from passing that location around 50mph and 31545 passed Chatham back on schedule – at least from Herne Bay – averaging 51mph between Newington and the Gillingham slowing. The 'D1' then held 27½ mph on the four miles, mainly 1 in 100, of Sole Street bank, regaining two of the earlier lost minutes and then achieved 72mph in the dip after Fawkham before a succession of bad signal checks approaching and after Swanley. These cost six minutes at least and with more signal checks after Bickley, the train arrived ten minutes late at Bromley South where the recorder alighted. However, the net time from Herne Bay to Bromley was only just over 70 minutes against the 81 minutes schedule.

31505 on 11 August had an even heavier load of 350 tons gross and lost eight minutes in the initial section to Herne Bay, suffering some early signal checks and falling to the low twenties at the summit at MP 64¼. It lost a further two minutes to Faversham, fell to 27mph after Faversham and could barely raise 50mph on the undulating stretch to Sittingbourne, just averaging 50mph on to Gillingham, and dropping three and a half minutes to passing Chatham where the train was now eleven minutes late. 31505 managed to average 22½ mph on the climb to Sole Street, regaining a minute and reached 65mph through Farningham Road before the inevitable signal checks at Swanley and Bickley. Net time for the 81 minutes schedule from Herne Bay to Bromley South (51.8 difficult miles) was 80 minutes, and again the recorder left the train at Bromley now 13½ minutes late. The 'D1s' could still be sprightly with smaller loads but their main line work was now restricted to these heavily loaded and frequently checked summer Saturday trains.

In the last timetable before the Kent Coast electrification – the summer of 1958 – 31145, 31247, 31487, 31505, 31545 and 31735 as well as a handful of 'E1s' appeared regularly on summer Saturday relief expresses and acquitted themselves reasonably well despite their age. However, when the electrics appeared in June 1959, engines were either withdrawn, or the lower mileage engines transferred to Nine Elms although many immediately went into store at Feltham. 31246, 31494, 31509, 31545 and 31727 were used occasionally on local freights or empty stock trains, with an occasional sortie on a Waterloo-Basingstoke semi-fast. In 1960/1 there was some goods and parcels train working in the Eastleigh area. 31735, observed on such a working, was withdrawn in March 1961 having amassed almost two million miles since first constructed as a 'D' in January 1902.

Just six remained on the Eastern and Central Sections after June 1959, 31487, 31489 and 31492 working from Tonbridge to Brighton and Eastbourne, and also on the Oxted trains. 31739, 31743 and 31749 worked out of Bricklayers Arms from Holborn Viaduct and London Bridge, including the 7.24am London Bridge-Ramsgate, a turn they could be found on regularly when all other Eastern Section 4-4-0 services (apart from the 'Schools') had ceased. 31739 was the last 'D1' to receive a significant ('Heavy Intermediate') repair at Ashford Works in September 1959 and regularly appeared on the 9.41am Brighton-London Bridge

7.24am London Bridge-Ashford /Ramsgate, 7.9.1960
31487 – Tonbridge
8 vehicles, 173/195 tons
(Timed by C.Hudson)

Miles	Location	Times	Speeds	Punctuality
0.0	London Bridge	00.00		1 L
3.0	New Cross	06.18	29/52	2 L
3.75	St Johns	07.15	47	
5.35	Hither Green	09.29	39/47	2 L
7.15	Grove Park	12.09	40/45/42	
8.4	Elmstead Woods	14.12	37/40	
9.4	Chislehurst	15.38	45	¾ L
10.8	Petts Wood	17.47	42/sigs	
11.95	Orpington	20.02		1 E
0.0		00.00		½ E
1.5	Chelsfield	04.02	38	
2.75	Knockholt	06.18	33/39/35	
5.45	Polhill	09.50	60	
6.75	Dunton Green	11.16	62	
8.3	Sevenoaks	13.26		T
0.0		00.00		½ L
0.6	Sevenoaks Tunnel N	02.15	29	
4.95	Hildenborough	07.34	65/sigs	
7.45	Tonbridge	12.47	(11 ½ net)	2¼ L
0.0		00.00		2 L
5.3	Paddock Wood	08.12	67	2¼ L
9.85	Marden	12.52	58/63	
12.35	Staplehurst	15.19	64/ sigs	
15.7	Headcorn	21.21	33*/sig10*	
20.9	Pluckley	28.37	43/57/48	
26.55	Ashford	36.33		6½ L

parcels train. After a few special workings (in common with 'E1' 31067) in 1961 it was withdrawn in November having achieved a mileage of 2,002, 974 – the only 'D1' to exceed two million miles in traffic.

Two Maunsell 4-4-0 rebuilds, 'D1' 31749, and 'E1' 31067, joined together for a final run with a train of wagons from Bat & Ball to Ashford, en route for withdrawal there. David Clark got wind of the planned final run and photographed the pair at several times on their melancholy last journey. Unfortunately, neither were saved from the scrapheap.

31727 with a local train at Ashford, c1958.
J.M. Bentley Collection

31489 with a Tonbridge-Redhill-Guildford train at Bough Beech, c1959.
Ken Wightman

31739 departs Ramsgate with a London bound van train, passing a 'C' goods engine in the sidings, 22 May 1959.
MLS Collection

31487 departs Tonbridge with a local service for Ashford, 27 August 1960.
David Clark

31739 on the 7.24am London Bridge-Ramsgate, the last regular passenger turn for a 'D1' or 'E1' on the Southern Region's Eastern Section. It is seen here passing Chislehurst, 2 August 1960. David Clark

31739, one of the last active 'D1s', at Petts Wood with the lightweight empty stock off the 5.45am London Bridge-Hastings, 30 June 1960. David Clark

31739 pilots 'E1' 31067 on a special train to the Hawkhurst branch, seen here at Paddock Wood and Cranbrook, 28 May 1961. David Clark

31749 and 31067 at Knockholt working to Bat & Ball where they will pick up a raft of wagons en route to store, withdrawal and scrapping at Ashford, 4 November 1961.
David Clark

31749 and 31067 run a train of wagons from Bat & Ball to Ashford on their final day in traffic, 4 November 1961.
David Clark

The preserved 737, built in 1901 at Ashford Works, restored to 1901 condition ready for display at the National Railway Museum, York, May 1960. G. Barlow/ J.M. Bentley Collection

Unrebuilt 'D', 31075, built by Dübs in March 1903, at Guildford on a Reading-Redhill stopping train composed of a 3-car BR Mark 1 set. I experienced this engine several times both on the 2.23pm Guildford at Wanborough and later hauled by it between Guildford and Reading a couple of times whilst making my way to Oxford to sit college scholarship attempts. 31075 was the last survivor of the unrebuilt 'Ds', being withdrawn at the end of 1956. David Maidment

Preservation

Unfortunately, no examples of either the 'D1' or 'E1' 4-4-0s have been preserved, but one of the original 'D' 4-4-0s in its full glory of South Eastern & Chatham Railway splendour as built in 1901 has been fully restored and is a prized possession of the national collection and displayed at the National Rail Museum in York.

Personal Reminiscences of the 'D1s'

My recollections of the 'D1s' are even more sketchy than my memory of the 'Ls'. I did better with the un-rebuilt 'Ds'. My 1951 *Southern ABC* indicates that I'd seen four 'Ds', and a few 'E1s' but not a single 'D1'. Perhaps this was because the 'E1s' at that time were Stewarts Lane based, while the 'D1s' were more dispersed to Kent Coast sheds or Bricklayers Arms. My school days at nearby Charterhouse enabled me to frequent the Guildford-Reading line as an observer of them at Wanborough, as a photographer there and at Ash Junction and actually riding behind them to Reading to sit (unsuccessfully) scholarship exams at Oxford. I could cycle to Wanborough after lunch on winter days when the afternoon was free until 4.30pm lessons, and the 2.23pm Guildford-Reading was a regular for a Guildford or Redhill 'D' – most often 31577 or 31586, although others did appear. I photographed 31586 at Ash Junction on this train and rode behind 31075 twice in my sorties to Reading and Oxford and returned once behind 31488.

The 'D1s' did not venture to this line, at least not west of Redhill, in the 1950s. I did try to get a run behind one of them in 1959. My meagre college grant would occasionally permit me to go to Victoria and buy a return ticket to Bromley South, rather than use my grant paid-for season ticket to Woking. One aim was to get runs behind the Eastern Section six-wheeled tender 'Arthurs' (my specialist subject was mediaeval German Arthurian epics involving such characters as Sir Ector de Maris, Sir Dodinas le Savage and Sir Harry le Fise Lake). A second and equal priority was to get a run or runs behind 'E1s' or 'D1s'. I did rather better at the first, encountering 30797, 30800, 30803 and 30805. I got one solitary run behind a 'D1', when 31505 turned up one lunchtime after I'd ventured out to Bromley on the 1.35pm Victoria-Ramsgate behind a 3-cylinder 'U1'. Regrettably, I took no notes of the journey, hanging out in the corridor near the rear of the train, enjoying the frequent sight of our 4-4-0 wending its way round the numerous curves en route to Victoria. I have no idea if we were on time – I don't think I even knew what the train was. I just waited at Bromley until something interesting turned up! One other experience that hardly counted was to be banked one day to Grosvenor Bridge on one of my Bromley South trips by 31545, which had brought the empty stock into Victoria.

I have to report one 'near-miss'. I decided to splash out on a return ticket to Chatham on the penultimate day of steam on the former LC&DR route in the hope of getting a 4-4-0. I arrived at Victoria at noon and seeing a 'new' (for me) 'Schools' on the 12.35 to Ramsgate (30914), decided to take that and see what I could pick up on the return. I watched rather enviously as 30803 roared through Chatham non-stop in the down direction shortly afterwards and settled down to a succession of Bulleid pacifics and Standard 5s in the up direction. Eventually I got an 'L1' (see Chapter 6) and was relieved and quite pleased. However, I didn't know what I'd missed until I was researching for this book. I not only discovered a log of 'D1' 31749 on the 11.50am Victoria-Dover (which I'd therefore missed by ten minutes) but also found a colour photo of it leaving Newington in the collection of Dick Riley's slides loaned to me by Rodney Lissenden. The only consolation is that the train had an appalling delayed journey, culminating in a series of crawls from signal to signal from the foot of Sole Street bank into Chatham station, reached over half an hour late.

31545 at Beckenham Junction with a Victoria-Margate train, 8 July 1956. R.C. Riley

31545 descends Sole Street bank with a Victoria-Margate train, c1957. Ken Wightman

31749 with the empty stock of a Margate train at Bromley South, 23 June 1957. Ken Wightman

31741 with a train of vans from Ramsgate at Shortlands Junction, October 1958.
Ken Wightman

31749 at Newington on the 11.50am Victoria-Dover, 13 June 1959 (the last Saturday of the timetable before the start of the electric trains on the following Monday). I missed this train by ten minutes, travelling on the following 12.35pm Victoria-Ramsgate train with a 'Schools'. This run was timed by both S.A.W. Harvey and D. Twibell but was ruined by signal checks throughout. Comments were made that the engine blew off steam frequently. It managed 68mph at Farningham Road after a severe p-way restriction at Swanley but took over half an hour to get from the foot of Sole Street bank to Chatham, spending several minutes at a number of signal stops all the way to Chatham. R.C. Riley

The fireman of 31489 hauls down coal to the front of the tender at Ashford on the 7.24am London Bridge-Ramsgate, while a BR Standard 4 tank busies itself in the sidings with empty stock, 14 May 1960. R.C. Riley

31509 and a row of stored 'D1s' at Feltham, 16 August 1959. R.C. Riley

31739 and 'E1' 31067 at Goudhurst on the special train to Benenden Girls' School on the Hawkhurst branch, 28 May 1961.
R.C. Riley

The final run of 31749 and 31067 on a freight to Ashford before withdrawal and scrapping was also covered in colour by Dick Riley. The pair are seen here at Bat & Ball passing a Bulleid suburban electric train unit, 4 November 1961.
R.C. Riley

Chapter 6

THE 'L1' DESIGN & CONSTRUCTION

The South Eastern Railway introduced a timetable which included some 80-minute non-stop schedules between Charing Cross and Folkestone in 1922. The load, normally restricted to around 220 tons, included Pullman cars and initially the trains were rostered to be hauled by the 'L' 4-4-0s. However, although they were fine for the heavily graded parts of the route, especially at the London end, they fell short in practice over the Tonbridge-Ashford section where speeds of the mid-seventies were required to keep time. Losses took place regularly as average speeds of around 60-62mph only were being achieved with top speeds in the 68-70mph range. Trials between a 'D1', the 'K' 2-6-4T and an 'L' took place and the 'D1' seems to have demonstrated that it could be the ideal engine for this, but the Maunsell rebuilds could not be spared from the Chatham route where the 'Ls', because of their heavier axleload, could not be used. Although the 'K' 790 performed well, there were doubts about its water capacity on a regular basis and Maunsell therefore sought a quick remedy, by considering whether the 'L' could be modified in a similar way to the 'Es' and 'Ds'.

As a temporary measure to help alleviate the complaints of poor timekeeping on the new expresses, some Drummond 'L12' 4-4-0s were drafted in after the Grouping for a few months and an 'L', A761, was modified in 1924 by adjustment of the rocking-lever arms and eccentrics to allow valve travel of 5³/₈in and increase the lap from ⁷/₈in to 1 ³/16in. It is surprising that this modification was not authorised for the rest of the class as this engine was recognised as being more free-running than the remaining engines of the class, but Maunsell decided on a new build of fifteen locomotives designed from the 'L' drawings, but with smaller cylinders, a higher boiler pressure, Maunsell superheater and improved cab. They were designated 'improved Ls' and thus the new class became officially identified as 'L1s'. Ashford couldn't undertake the work, so the North British Locomotive Company carried out the order at a cost of £5,925 per locomotive (they were said to have made a considerable loss on the contract – and barely made a profit on the later substantial order of the 'Scotchman N15s'). The fifteen engines, numbered A753-A759 and A782-A789, were delivered in March and April 1926.

Dimensions of the 'L1s' were:

Cylinders (2 inside)	19½in x 26in
Coupled wheel diameter	6ft 8in
Bogie wheel diameter	3ft 7in
Boiler pressure	180lbs psi
Grate area	22½sqft
Heating surface	1,642sqft
Axleload	19 tons 10 cwt
Locomotive weight	57 tons 16 cwt
Tender weight	40 tons 10 cwt
Tractive effort	18,910lbs
Tender capacity	
Water	3,500 gallons
Coal	5 tons

Use was made as far as possible of the drawings of the Wainwright 'L' class in order to expedite the construction of the engines that were urgently needed. Lessons had been learned now of the advantages of valve and motion arrangements as developed by Churchward and adopted by Maunsell for his earlier 4-4-0 rebuilds, but there was a limit to the improvements that could be made without a complete redesign of the cylinders. Valve travel was therefore 5³/₈in in full gear and lap of 1³/₁₆in as for the experiment on A761. The smokebox was similar to that fitted to the 'N' moguls and the chimney in outline similar to the 'U1' 3-cylinder 2-6-0s.

Livery was lined Maunsell green, with numbers painted in large numerals on the tender. In 1931, as with the former South Eastern locomotives, they had 1,000 added to become 1753-1759 and 1782-1789. In 1939 the lining was omitted, and Bulleid lettering became standard along with the number being painted on the cabside rather than tender. Then during the war plain black was adopted for the class. In May

A759 in Works Grey as delivered by the North British Locomotive Company, March 1926. F. Moore/MLS Collection

A755 as newly built,1926. W.J. Reynolds/MLS Collection

A757 at Bricklayers Arms, 1926. F. Moore/MLS Collection

A783 as built, c1927. F. Moore/MLS Collection

1758 after the 1931 SR renumbering, c1932. John Scott-Morgan Collection

1782 in wartime plain black livery at Bricklayers Arms, 22 September 1947. J.D. Darby/MLS Collection

1946 Ashford began repainting engines in malachite green, starting with 1786. Finally, all 'L1s' except 1753 and 1782 received the malachite green livery until after nationalisation the BR mixed traffic livery of lined black. 1755 and 1789 ran for a while with a small 's' before the number until all adopted the BR numbering of 31753-9 and 31782-9. The first to receive the BR livery was 31788 in January 1949. The last engines to retain the malachite livery with BR numbers were 31753/4/6 and 31785/6/9 which were repainted in 1953.

The final heavy repairs, with impending electrification spelling out the doubtful future of the class, were undertaken for 31753/7/9 in 1957, 31754 in 1958 and finally 31756 and 31786 in 1959. After electrification of the former LC&DR route in June 1959, 31755 and 31758 were condemned and 31757, 31759 and 31782 were stored. The remaining 'L1s' were transferred to Nine Elms for some relatively minor roles including parcels and empty stock working. 31784, 31785 and 31788 were withdrawn in early 1960 and the remainder placed in store except for 31753. Three more – 31754, 31756 and 31786 – were brought back for Christmas parcel train working in December 1960 and 31753, 31756 and 31786 were active in the summer of 1961. All remaining members of the class, including those in store, were withdrawn at the commencement of the winter timetable in 1961 apart

s1755 as numbered initially at nationalisation and in malachite green livery, 1949. J.M. Bentley Collection

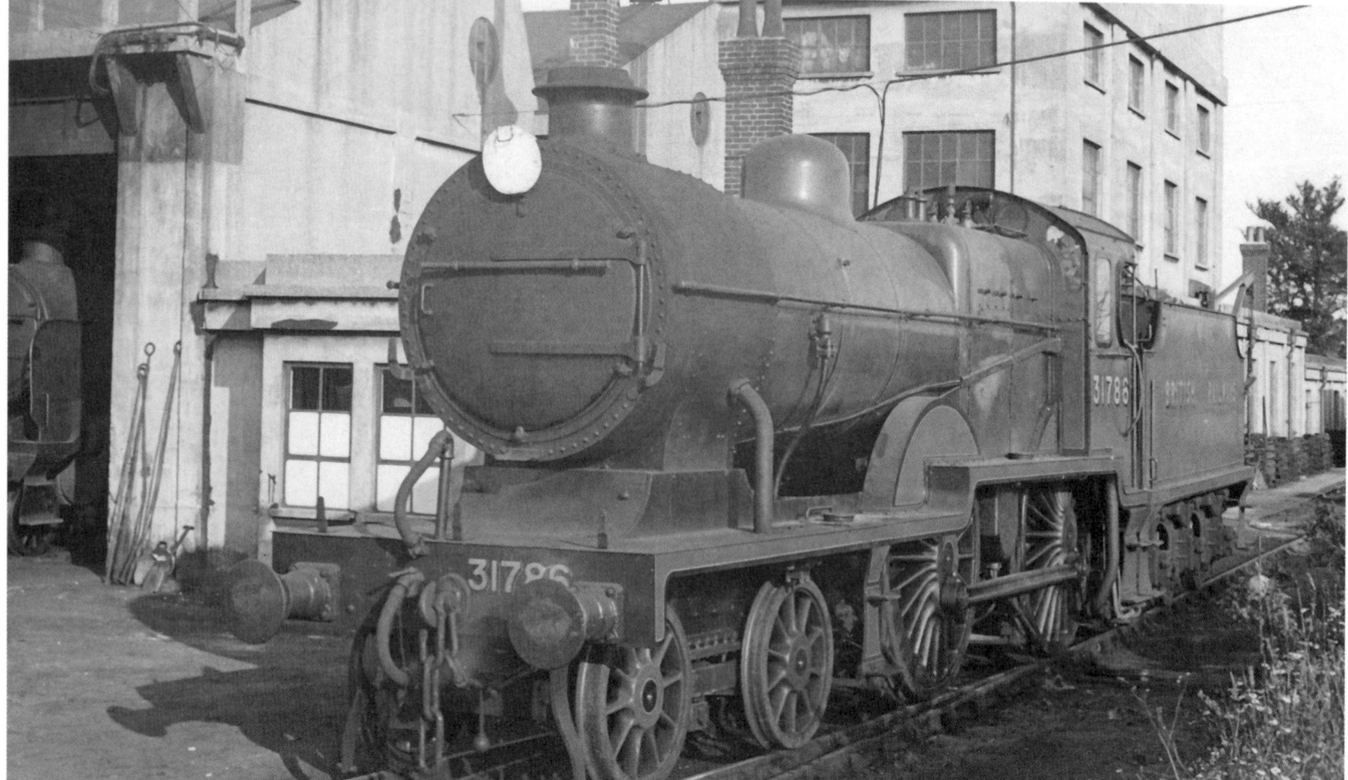

31786 in malachite green but repainted with BR number and 'British Railways' lettering on the tender, c1951. G. Aston/J.M. Bentley Collection

31787 still in malachite green but with BR numbering and lettering, Ramsgate, 2 July 1949. H.C. Casserley/J.M. Bentley Collection

from 31786 which eked out a few more miles before the class became extinct in February 1962, with a final mileage of 1,068,774. 31753 had attained the highest mileage of the class – 1,124, 008 and 31755 had also exceeded a million miles in traffic, despite being one of the earliest to be condemned. The engines were broken up at Ashford and Eastleigh Works, none going to private scrap yards, and unfortunately none were preserved, so that the heritage industry has no representative of the successful SR 4-4-0s except

31757 at Ashford, 7 June 1952. J.D. Darby/MLS Collection

31757 at Faversham, 30 June 1957. MLS Collection

31788 at Stewarts Lane being coupled to a Bulleid light pacific, possibly for haulage of the *Night Ferry*, 18 February 1958. MLS Collection

31754 ex-works at Ashford in company with 'H' 0-4-4T 31266, April 1958. J.M. Bentley Collection

for the very different and more powerful 'Schools' to be described in the next chapter.

Operation

As soon as the fifteen new locomotives were received from the North British Company in March and April 1926, they were rostered to the 80-minute Folkestone expresses and were found capable of timekeeping with up to 350 tons – a significant improvement on the 'Ls' which were hard pressed to keep time with more than 250 tons. The 'L1s' had a turn of speed that the 'Ls' found elusive and the *Railway Magazine* in the late 1920s published runs in the up direction of the Folkestone 80-minute trains which drew attention to this, quoting five runs with 753, 755, 758, 783 and 784 which averaged between 68.4 and 72.0mph for the 25.6 miles between Smeeth and Paddock Wood, with almost total uniformity of maximum speed – ranging from 74 to 76½mph (in fact the highest speed and average was with 758 and the heaviest load – 340 tons gross). They were also used on Hastings and Ramsgate expresses. The initial allocation was:

Dover: A753, 754, 756, 782-785
Bricklayers Arms: A755, 757-759, 786-789

In the period between 1922 and 1932, a regular traveller kept detailed records of running, and his analysis is illuminating, demonstrating the gulf in regular performance of the 'Ls' and the so-called 'improved Ls' – i.e. the 'L1s'. The average speed of 27 runs with 'Ls' between Ashford and Tonbridge was just 61.4mph with 275 tons, whereas the 'L1s' averaged 65.4mph over 32 runs and a heavier average load of 320 tons. This was similar to the performance of the 'L12s' and only just below that of the later more powerful 'Schools'. The first ten 'Schools' allocated to Dover and Eastbourne did not displace the 'L1s' from this work until more were delivered in 1932 and then, in January 1933, they were all congregated at Bricklayers Arms for semi-fast services over both the Kent Coast routes via Tonbridge and Chatham, working turn and turn about with the 'E1s' and 'D1s'.

Cecil J. Allen in his regular articles on 'British Locomotive Performance and Practice' in the *Railway Magazines* of 1932 and 1934 was primarily concerned with reporting performance of the new 'Schools' (see Chapter 7) on the Folkestone-Charing Cross 80-minute schedules but included a couple of runs in each article by 'L1s' for comparative purposes. 1757 with nine coaches, 292/310 tons, would have kept time but for a p-way slack after Orpington and signal delays in from New Cross. After 45mph through Sandling Junction, 1757 accelerated to 73 through Ashford, maintained the upper 60s between Ashford and Tonbridge and after the 30mph slowing through Sevenoaks, touched 61mph at Dunton Green, cleared Knockholt summit at 46mph and reached 71 at Orpington just before the slack. 1755 with an extra coach started even better with 48mph at Sandling Junction and 77½ through Ashford, half a minute faster than the lighter train, touched 78 at Headcorn, fell to 33mph before Sevenoaks tunnel, an excellent 51 at Knockholt and 72½ at Orpington and, unchecked, although with easy running in from Chislehurst, reached Charing Cross in 79 minutes 20 seconds, ¾ minute early. A new 'Schools' (903), also with ten coaches on the same train, was only a net minute faster, mainly between Staplehurst and Paddock Wood and a rapid descent from Chislehurst through Grove

A782 on a Dover-Charing Cross semi-fast train at Folkestone Warren, c1930. John Scott-Morgan Collection

Park – and that was the fastest of the heavier trains quoted.

In the May 1934 article, 1757 features again with an actual time of 80 minutes 5 seconds with the usual nine coach load with very similar speeds to the earlier runs, but 1758 with ten coaches, 320/340 tons excelled itself between Ashford and Tonbridge. It struggled to Ashford losing two minutes on the schedule, but once over the top of the bank at Westenhanger it flew and I show below an excerpt of the log.

Ashford-Tonbridge, c1933
1758
10 chs, 320/340 tons

Miles	Location	Times	Speeds	Punctuality
0.0	Ashford (pass)	00.00	76	2½ L
	Chart Sidings	01.42	71	
5.7	Pluckley	04.31	78	
10.9	Headcorn	08.35	75	
14.2	Staplehurst	11.09	78½	
16.7	Marden	13.11	73	
21.3	Paddock Wood	16.51	75	1 L
26.6	Tonbridge (pass)	24.36	sig stand/28*	4 L

1758 averaged 75.5mph between Ashford and Paddock Wood. After the signal check 1758 and its driver proceeded more normally with 46 minimum at Knockholt and 73 at Orpington before the run was ruined by signal checks in from Hither Green, arriving at London Bridge 6½ minutes late. A couple of runs on the 9.15am Charing Cross-Folkestone were timed by S.A.W. Harvey, the summer schedule requiring stops at Tonbridge and Ashford, but the winter 1933 plan had the 80-minute schedule with just one stop at Waterloo East. By this time the booked locomotive would have been a 'Schools' but the two runs quoted below show that the 'L1s' were quite capable of maintaining the schedule, given a reasonably clear road.

Charing Cross-Folkestone Central

		9.15am Charing Cross – Deal, 26.12.1933 1788 333/350 tons			**9.15am Charing Cross – Deal, 7.7.1933 1757 247/270 tons**		
Miles	**Location**	**Times**	**Speeds**	**Punctuality**	**Times**	**Speeds**	**Punctuality**
0.0	Charing Cross	00.00		T	00.00		T
0.8	Waterloo East	02.49		¾ L	02.35	30	½ L
		00.00					
1.9	London Bridge	03.10	30	¾ E	05.20	sigs	¾ E
4.9	New Cross	07.23	52	¾ E	12.30	52	T
5.6	St Johns	08.15	56		13.27	54	
7.2	Hither Green	10.10	50	¾ E	15.35	48	½ E
10.3	Elmstead Woods	15.25	40/33		20.43	40/38	
11.3	Chislehurst	17.15	44	¾ L	22.22	45	¾ E
13.8	Orpington	21.15	46/52	1¾ L	26.10	44/52	¼ L
15.3	Chelsfield	23.15	48		28.10	46	
16.6	Knockholt	25.10	46		30.10	40	
20.6	Dunton Green	29.30	74		34.50	70	
22.1	Sevenoaks	30.57	60	2 L	36.20	60	¼ L
27.0	Hildenborough	35.27	88		41.30	72	
29.5	Tonbridge	37.53	52*	1¾ L	47.20 (45 net)	sigs stand	2¼ L
					00.00		
34.8	Paddock Wood	43.30	70	1 L	07.45	70	¼ E
39.4	Marden	47.25	64		12.18	62	
41.9	Staplehurst	49.45	74		14.50	70	¼ E
45.2	Headcorn	52.40	72		17.55	70	
50.4	Pluckley	57.30	70		23.30	58	
56.1	Ashford	62.55	74	1 L	30.17		¾ E
					00.00		T
60.4	Smeeth	67.05	62		07.00	58	
64.2	Westenhanger	71.10	60		11.05	64	
65.4	Sandling Junction	72.35	64	½ L	12.20	64	¼ L
68.7	Cheriton	75.50	72		17.20	sigs 36*/43	
69.2	Shorncliffe	76.16	66		18.10	sigs 5*	
70.7	Folkestone Central	77.28		½ L	20.55	(17 net)	4 L

1788 with the heavy load, after the excellent run from London, found the short start to stop times onto Deal too tight and dropped four minutes to Martin Mill, then recovered slightly and arrived at Deal 3 ¼ minutes late. 1757's run was more routine but received more signal checks at Folkestone Warren and approaching Deal and arrived there eight minutes late.

Now for three runs, two on consecutive days in 1934 and one the following year on the LC&DR route on the 5.22pm Cannon Street to Dover, timed in detail as far as Chatham (stops at Gillingham, Sittingbourne, Faversham, Selling, Canterbury East and all stations to Dover Priory made the train less interesting from a performance point of view after that).

		5.22pm Cannon Street-Chatham (Dover)								
		16.8.1934			**17.8.1934**			**16.8.1935**		
		1757			**1755**			**1755**		
		216/236 tons			**216/240 tons**			**220/240 tons**		
Miles	**Location**	**Times**	**Speeds**		**Times**	**Speeds**		**Times**	**Speeds**	
0.0	Cannon Street	00.00		T	00.00		T	00.00		T
0.7	London Bridge	02.20		¼ E	02.20		¼ E	02.20		¼ E
3.7	New Cross	10.00	sigs stand		07.05	48	T	07.38	sigs	¾ L
4.4	St Johns	11.20	40		08.05	50		08.42		
6.0	Hither Green	13.50	45	3¼ L	10.30	46	T	11.25	34/44	1 L
9.1	Elmstead Woods	18.37	42/40		15.27	43/44		15.52	42	
10.1	Chislehurst	20.00	46	3½ L	16.55	48	½ L	17.10	46	¾ L
12.4	St Mary Cray	23.40	62		20.42	66		20.45	68	
15.3	Swanley Junction	26.45	54	3¼ L	23.52	54	¼ L	23.43	60	¼ L
18.1	Farningham Road	29.20	75		26.32	80		26.15	76	
21.0	Fawkham	31.53	56		29.10	64		28.46	62	
23.5	Meopham	34.45	56		32.10	56		31.34	57	
24.5	Sole Street	35.55	56/65	1¾ L	33.22	50/62	¾ E	32.40	52/60	1¼ E
28.5	Cuxton Road	40.00	50*		37.27	60	1½ E	37.00	50*	2 E
30.5	Rochester Bridge	43.22	30*	¾ L	40.43	30*	1¾ E	40.17	34*	2¼ E
31.3	Rochester	44.58 (42 net)		1 L	42.30		1½ E	41.52 (40 ½ net)		2¼ E
0.0		00.00			00.00			00.00		
0.6	Chatham	02.33		1½ L	02.32		½ L	02.39		¾ L

1755 on a Charing Cross-Hastings train leaves Sevenoaks tunnel, c1933. John Scott-Morgan Collection

1756 with a Charing Cross-Folkestone express at Chelsfield, c1934.
J.M. Bentley Collection

1756 with the Royal Train made up of Pullman cars en route for Ascot, the Derby, June 1938.
J.M. Bentley Collection

1757 lost nearly three minutes after a severe signal check outside Sittingbourne but held time thereafter. 1755 in 1934 was also checked outside Sittingbourne but regained it before Dover. The recorder (S.A.W. Harvey) left the train at Canterbury East in 1935, where it was a couple of minutes late after a pws to 15mph between Selling and Canterbury.

The 'L1s' were economical engines and their maintenance was relatively light compared with the older machines – between 1933 and 1936 they averaged over 76,000 miles between Works repairs, only surpassed by the even newer 'Schools'. In 1930 1753-1759 were transferred from Bricklayers Arms to Dover. They continued on semi-fast services on both Kent Coast routes throughout the 1930s, being particularly active on holiday trains on summer Saturdays, and were also often used on additional excursion trains to the South Coast resorts such as Eastbourne and Bognor, or even to Bournemouth via the Central Section and Southampton. They were also to be found on relief boat trains. In 1936 the *Night Ferry* service was introduced and the Pullman sleeping car express was found to be too heavy for the 'Lord Nelson' 4-6-0s and was rostered instead for a couple of 4-4-0s, usually including at least one 'L1'. There were also some Margate-London services via Redhill, as well as cross-country trains from the Kent Coast to the West Midlands via Redhill.

Priorities changed at the onset of the Second World War and several 'L1s' were put in store in the autumn of 1939. However, winter parcels traffic and then the Dunkirk evacuation saw all them back in traffic, with the 4-4-0s working from Dover or Folkestone to Kensington or Reading for onward movement by the LMS or GWR.

Mr Harvey continued to record his journeys during the Second World War, some of which give evidence of the disruption caused by enemy action, others reveal near peace time operation. Below is one of the latter.

Tonbridge-London Bridge, 22.7.1941
3.24pm Tonbridge (ex-Ashford)
1786
259/275 tons
Driver Perry, Bricklayers Arms

Miles	Location	Times	Speeds	Punctuality
0.0	Tonbridge	00.00		T
2.5	Hildenborough	05.20	36	
4.8	Sevenoaks Tunnel	-	32/36	
7.4	Sevenoaks	14.21		¾ E
0.0		00.00		
1.5	Dunton Green	02.57		
2.8	Polhill	04.44		
5.5	Knockholt	08.18	45	
6.8	Chelsfield	09.42	60	
8.3	Orpington	11.10	65	¾ E
9.5	Petts Wood	14.50	pws 15*	
10.8	Chislehurst	16.10		1¼ L
11.8	Elmstead Woods	17.13	60	
13.1	Grove Park	18.30	63	
14.9	Hither Green	21.30	pws 15*	1½ L
16.5	St Johns	24.05		
17.2	New Cross	25.17		2¼ L
21.0	London Bridge	29.48	(25 net)	¾ L

With a number of South Eastern Section locomotives moved to the Western Section to replace Southern 4-6-0s required on other companies for heavy wartime freight traffic, the 'L1s' found themselves with plenty of work to cover in Kent. The lack of maintenance meant that the simple 2-cylinder 4-4-0s were more suited to the conditions than the 3-cylinder 4-4-0s and 2-6-0s. For most of the war, the 'L1s' were based at Bricklayers Arms, where despite heavy bombing and air attacks, they stayed relatively unscathed. Periods between heavy repairs were extended with some locomotives exceeding 100,000 miles, although as can

31782 passes Graveney Sidings with a Ramsgate-Nottingham summer Saturday returning holiday express, 28 July 1951. Ken Wightman

31786 with a London Bridge-Eastbourne excursion, July 1951. J.M. Bentley Collection

be imagined, wear and tear was endemic, and the engines emerged from the war in 1945 in a very rundown condition. Despite this they were required then to handle many of the heavy service trains of troops returning from the continent, working to the North and West Midlands via Kensington and Reading. 1753-1756 were transferred to Dover for some of this business.

After the war, and essential repair work, they resumed service on Kent Coast semi-fast trains and, as holiday traffic returned, on summer Saturday relief expresses. S.A.W. Harvey was continuing to record performance and the run to Chatham below was well up to pre-war standards.

Cannon Street-Chatham, 22.3.1948
4.32pm Cannon Street
1788 – Bricklayers Arms
240/255 tons

Miles	Location	Times	Speeds	Punctuality
0.0	Cannon Street	00.00		T
0.7	London Bridge	02.41		¼ E
0.0		00.00		
3.0	New Cross	04.55		T
3.7	St Johns	05.45	sigs	
5.3	Hither Green	08.30		T
8.4	Elmstead Woods	13.15	44	
9.4	Chislehurst	14.40	45	¾ E
11.8	St Mary Cray	17.38	46/65	
14.4	Swanley	20.38	45	2¼ E
17.5	Farningham Road	23.52	74	
20.4	Fawkham	26.30	65	
22.9	Meopham	29.15	58	
23.9	Sole Street	30.48	sigs	2¼ E
27.9	Cuxton Road	35.48	sigs 48*	2¼ E
29.9	Rochester Bridge	38.58	30*	2½ E
30.7	Rochester	40.32	(36 ¾ net)	2½ E
0.0		00.00		
0.6	Chatham	02.47		¾ L

s1789 as renumbered and lettered 'British Railways' shortly after nationalisation on a Charing Cross-Hastings train at Crowhurst, August 1948. J.M. Bentley Collection

In April 1950 the 'L1' allocation was:

Dover: 31753, 31754, 31757, 31759
Bricklayers Arms: 31755, 31758, 31785, 31788, 31789
In store: 31756, 31782-31784, 31786, 31787

The stored engines were put back into traffic for the summer service.

Next, therefore, we have two examples of the work that the 'L1s' performed on post-war summer Saturday holiday expresses, both returning from the Kent Coast, stopping to pick up at stations from Ramsgate to Whitstable and then running fast calling at Gillingham, Chatham and Bromley South only. Both were excellent efforts, arriving on time in London despite signal checks in the latter stages of the run. 31782 lost some time on the Whitstable-Gillingham section after a slow climb from Sittingbourne but displayed more energy after Chatham to pick up time. 31787 ran comfortably to time at all locations until nearly brought to a stand at Beckenham Junction. This locomotive had only been taken out of store for the 1952 summer season.

		4.18pm Herne Bay-Victoria (from Ramsgate), 30.8.1952 **31787 – Bricklayers Arms** **265/290 tons**			**4.18pm Herne Bay-Victoria, 11.7.1953** **31782 – Bricklayers Arms** **260/280 tons**		
Miles	**Location**	**Times**	**Speeds**	**Punctuality**	**Times**	**Speeds**	**Punctuality**
0.0	Herne Bay	00.00		T	00.00		T
2.2	Chestfield	04.46		¼ E	04.59		T
		00.00			00.00		
1.4	Whitstable	03.53		T	03.56		T
		00.00			00.00		
4.0	Graveney Sidings	06.48	64		06.49	65	
7.1	Faversham	10.30		½ L	10.24		½ L
9.1	MP 50 (summit)	13.14	49		13.11	48	
11.1	Teynham	15.40	68		15.47	66	
14.4	Sittingbourne	18.55	64	1 L	19.25	55	1½ L
16.4	MP 42 ¾ (summit)	21.25	45		22.33	38	
17.5	Newington	22.45	59		24.05	54	
20.2	Rainham	25.38	67		27.10	60	
23.2	Gillingham	29.33		½ L	32.17		3¼ L
		00.00			00.00		
1.6	Chatham	03.50		T	03.40	50	3L
		00.00			00.00		
2.2	Rochester	04.42			04.30		
3.0	Rochester Bridge	06.00		½ E	05.35	48	2 L
5.0	Cuxton Road	09.58			09.02	30	
9.0	Sole Street	19.27	30/26	½ E	19.10	23	1¼ L
10.0	Meopham	21.04			20.56		
12.5	Fawkham	24.01	55		23.50	60	
15.4	Farningham Road	26.48	70		26.24	75	
18.5	Swanley	30.11		¼ L	29.38	48	2½ L
21.1	St Mary Cray	32.57	68		32.32	65	
23.9	Bickley Junction	36.25	sigs	½ L	35.08		2 L
25.0	Bromley South	39.05		T	37.31		1½ L
		00.00			00.00		
0.8	Shortlands Junction	02.30			02.05		
2.2	Beckenham Junction	07.00	sigs 5*	3 L	04.44	sigs	2¼ L
3.6	Penge East	10.50			06.57		
5.1	Sydenham Hill	13.50	30		09.52	32	
6.9	Herne Hill	16.15	55	3¼ L	12.00	60	½ L
7.7	Brixton	17.34		2½ L	16.50	2 min sig stand	3¼ L
10.2	Grosvenor Bridge	21.40	sigs 10*		21.40		
10.9	Victoria	24.20	(20 net)	1¼ L *	23.32	(19 ½ net)	2 L**
		*2 E by public timetable			** 1 E by public timetable		

31782 with a Tonbridge-Ashford local train at Paddock Wood, 6 September 1952. MLS Collection

31783 at Crowhurst with a Charing Cross-Hastings express, July 1953. Rodney Lissenden Collection

In December 1952, 1786-1789 inclusive were transferred to the Southern Region's Western Section at Eastleigh depot to replace life-expired Drummond 4-4-0s on local passenger, parcels and freight work. Some aspects of their design were appreciated (particularly the less Spartan cab layout and protection) but they were not particularly suited to some of the work that involved tender first operations. Their 6ft 8in coupled wheels also were less appropriate for the work than the smaller wheeled Drummond engines and when the BR Standard '4' 2-6-0s became available in 1954, the 'L1s' were despatched back to the Eastern Section at Gillingham and Faversham. One of them, 1787, was reported in the August 1954 edition of the monthly *Trains Illustrated* magazine as working back to London en route to Kent on the 10.15am additional Lymington Pier train instead of the usual 'D15' or 'T9'. Apparently, the crew and engine put up a spirited performance running up to Waterloo in 92 minutes net from Southampton Central.

31756-31759 and 31782 were transferred to Ashford at the beginning of the winter timetable in 1953 and 31785-31787 went to Gillingham at the beginning of the summer timetable in 1955, before moving again to Faversham. In winter most were stored.

A few 'L1s' were still working on Dover-Victoria services prior to the electrification which commenced in June 1959. 31753 was in good form in April of that year on the 1.57pm from Dover Priory stopping at all stations to Faversham, where it arrived on time. The log on page 120 takes up the running from that point. The climb of Sole Street bank and subsequent speed (and the very early arrival at Victoria) was particularly noteworthy at this time.

An official BR publicity photo of 31785 piloting a Bulleid light pacific at Victoria before departure with the *Night Ferry* to Dover, Dunkirk and Paris, 1956. British Railways/ J.M. Bentley Collection

31753 on an up local stopping train at Folkestone Junction, 13 July 1954. MLS Collection

Faversham-Victoria, 18.4.1959
1.57pm Dover Priory-Victoria
31753 – Dover
Driver P.Tutt
230/245 tons

Miles	Location	Times	Speeds	Punctuality
0.0	Faversham	00.00		T
1.9	MP 50 (summit)	04.40		
4.0	Teynham	08.20		¼ L
0.0		00.00		
3.3	Sittingbourne	07.07	sigs	1 L
0.0		00.00		
2.0	MP 42 ¾	05.00		
3.1	Newington	06.50	pws	
5.8	Rainham	12.17		
8.8	Gillingham	17.42		¼ E
0.0		00.00		
1.6	Chatham	04.05		T
0.0		00.00		
1.4	Rochester Bridge	02.50		¼ L
3.4	Cuxton Road	06.17	36	
7.4	Sole Street	13.20	38	2¾ E
8.4	Meopham	14.41	52	
10.9	Fawkham	17.39	65	
13.8	Farningham Road	20.12	76	
16.9	Swanley	23.28	sigs	5½ E
19.5	St Mary Cray	27.57	pws	
21.6	Bickley Junction	31.20		4¾ E
23.4	Bromley South	35.04		4 E
0.0		00.00		4 E*
0.8	Shortlands Junction	01.53	pws	
2.2	Beckenham Junction	04.29	52/56	4½ E
3.6	Penge East	06.09	48	
5.1	Sydenham Hill	08.05		
6.9	Herne Hill	10.27		10 E
7.7	Brixton	11.38	50*	9½ E
8.6	Clapham	13.06		
10.2	Grosvenor Rd Bridge	16.18		
		17.30/17.40 sig stand		
10.9	Victoria	19.32	(17 ¾ net)	7 E

* Left Bromley South early (set down only)

31756 performs humble duties with a pick-up freight at Bexhill, September 1956. J.M. Bentley Collection

31753 with an up Dover train for Victoria at St Mary Cray during the track quadrupling ready for the imminent electrification, 2 May 1959. R.C. Riley/Transport Trust Collection

31786 leaving Wadhurst with a Tonbridge-Hastings stopping train, June 1957. Ken Wightman

31786 at Shortlands Junction with a summer Saturday Dover – Victoria train, 2 August 1958. R.C. Riley

31754 arriving at Faversham on the last Saturday of steam on the Chatham route to the Kent Coast, 13 June 1959. R.C. Riley

Personal Reminiscences of the 'L1s'

I had no experience of Maunsell's 'L1s' (apart from seeing them as a trainspotter in the early 1950s) until the penultimate day of the SR Eastern Section timetable before the Kent Coast electrification was implemented. I had gone down to Chatham, hoping for a run behind an 'E1' or 'D1', and, unknown to me at the time, had just missed 31749 on the 11.50am Victoria – see previous chapter. I compensated with a run behind a 'Schools' and waited for something more interesting on the way back. However, the up trains that Saturday afternoon were entirely in the hands of Ramsgate 'Battle of Britains' and Stewarts Lane's Standard 5s (73080-73089) and I had almost given up when a rundown looking 'L1', 31788, turned up and I gratefully accepted that. Despite appearances, the run was very respectable, and a log is shown on the next page – with additions from a couple of other train timers that were on the train from Canterbury East.

Canterbury East-Victoria, 13.6.1959
2.42pm Canterbury East (ex-Ramsgate due Victoria 4.37pm - I joined at Chatham)
31788
228/245 tons
(Timed S.A.W. Harvey)

Miles	Location	Times	Speeds	Schedule
0.0	Canterbury East	00.00	pws 20*	T
6.6	Selling	12.05		T
0.0		00.00		
3.2	Faversham	05.40		¼ E
0.0		00.00	55	
4.0	Teynham	07.45		¼ E
0.0		00.00		
3.3	Sittingbourne	05.55		T
0.0		00.00		
2.0	MP42 ¾	05.12	35	
3.1	Newington	06.55	54	
5.8	Rainham	10.05	63	
8.8	Gillingham	14.02		1 L
0.0		00.00		
1.6	Chatham	04.21		1¼ L
0.0		00.00		1 L
1.4	Rochester Bridge Jcn	03.18		1¾ L
3.4	Cuxton Road	07.05	39/32	
5.0	MP 5	-	36	
7.4	Sole Street	13.56	39	1 L
10.9	Fawkham	18.04	68	
13.8	Farningham Road	20.29	73	
16.8	Swanley	23.41	51/sigs 35*	4¼ E
19.5	St Mary Cray	26.44	53	
21.6	Bickley Junction	sig stand – 3 mins		
23.4	Bromley South	38.12	(31 ¼ mins net)	¼ L
0.0		00.00		
0.8	Shortlands Junction	01.50	28/32	
2.2	Beckenham Junction	04.30	pws 20*	¼ E
3.0	Kent House	06.45/07.52 sig stand		
3.6	Penge East	11.20		
5.1	Sydenham Hill	14.37		
6.9	Herne Hill	17.37	45/sigs 20*	1¾ L
7.7	Brixton	18.59	sigs	2¼ L
8.6	Clapham	20.30		
10.2	Grosvenor Rd Bridge	23.02		
10.9	Victoria	24.56	(17 ¾ mins net)	2¼ L

After the electrification of the Chatham route, several of the 'L1s' that were in good condition were transferred for a few months to Nine Elms and worked semi-fast trains to Basingstoke and parcels trains during the summer of 1959 and 1960, with 31786, the one most recently ex-works, surviving there still operational in 1961. My first encounter was with 31786, looking very smart, on a run from Farnborough to Woking after visiting the Railway Enthusiast Clubrooms at Farnborough. It had worked down on the Saturday 12.42 to Basingstoke and I picked it up on the 7.40pm from that town.

Farnborough-Woking
7.40pm Basingstoke-Waterloo, 25.9.1959
31786 – Nine Elms
6 chs + 1 van, est 220 tons

Location	Times	Speeds	Punctuality
Farnborough	00.00		T
MP 32	02.57	48	
MP 31	04.17	52	
MP 30	05.26	59	
MP 29	06.27	60	
Brookwood	08.07		T
	00.00		
MP 27	02.15	49	
MP 26	03.30	57	
MP 25	04.46	49 easy	
Woking West Junction	06.09/07.42 sigs stand		
Woking	09.04		2 L

31788 trails out of Victoria station, having arrived on the Ramsgate train due Victoria at 4.37pm on the last Saturday of steam operation of the former LC&DR route, a train on which I travelled from Chatham, 13 June 1959. David Maidment

31786 arrives at Woking on the 12.42pm Waterloo-Basingstoke, a train which it worked on successive Saturdays, returning to London on the 7.40pm Basingstoke on which I travelled from Farnborough to Woking, July 1959. David Maidment

I had a couple of runs from Woking to Waterloo, behind 31788 on the first up semi-fast (6.39am Basingstoke) in the summer of 1959, a surprise as it looked even more rundown than when I'd experienced it a month earlier at Chatham. However, it was sprightly enough running in the upper 60s before signal checks from Surbiton onwards. Then a year later, I was even more surprised to find 31753 still active from Nine Elms on the morning stopping train from Bournemouth that left Woking at 9.48am for a non-stop run from there to Waterloo, a train booked for a Nine Elms Standard 5 or 'King Arthur'. The log is shown on page 126.

31786 at Farnborough with the 12.42pm Waterloo-Basingstoke, July 1959. David Maidment

		Woking-Waterloo, 3.6.1960 **10.22am Waterloo arrival ex-Bournemouth (9.48am Woking)** **31753 – Nine Elms** **7 chs, 240 tons**			**2.9.61** **12.53pm Basingstoke-Waterloo** **31786 – Nine Elms** **6 chs, 197 tons**		
Miles	**Location**	**Times**	**Speeds**	**Punctuality**	**Times**	**Speeds**	**Punctuality**
0.0	Woking	00.00		T	00.00		¾ E
2.65	West Byfleet	04.16	61		04.17	65	
5.2	Weybridge	06.56	63/60		06.41	66/sigs 44*	
7.2	Walton	09.07	66		09.26	sigs 15*	
8.35	Hersham	-	68		11.17	35/49	
9.9	Esher	11.42	70		13.09	51	
10.95	Hampton Court Jcn	12.45	69		14.22	52	1¾ L
12.25	Surbiton	13.57	65		16.20	sigs 40*/55	
14.55	New Malden	16.15	64		19.04	62	
17.1	Wimbledon	18.48	65		21.43	66	
18.75	Earlsfield	20.30	68		23'19	67	
20.4	Clapham Junction	22.31	30*		25.16	42*	1¾ L
21.55	Queens Road	-	sigs 0*		27.18 sigs 34*/46		
23.0	Vauxhall	29.27	sigs 5*		29.26	41/sigs	
24.3	Waterloo	33.20	(28 ½ mins net)	2¼ L	34.13	(29 net)	3¾ L

The September 1961 run, just a few months before 31786's withdrawal, was timed from Basingstoke by Mr C. Hudson. Stopping at all stations to Woking except Brookwood, it ran ahead of its scheduled start-to-stop times without exceeding 50mph, except on the last stretch from Farnborough when it accelerated up to 75mph on the descent from MP 31 until almost halted by signals at Woking Junction.

Chapter 7

THE 'V' ('SCHOOLS') DESIGN & CONSTRUCTION

The principal reason for the design of the 'V' 'Schools' class 4-4-0 seems to have been the limited kinetic envelope and restricted clearance on the Tunbridge Wells-Hastings route. When constructed in 1852, the loading gauge was smaller than on other lines, particularly the double-line tunnel at Mountfield near Battle. Only since partial singling after electrification has the clearance problem been overcome.

In the 1920s, the Southern Railway was strengthening the former SE&CR routes to enable more powerful locomotives to handle boat trains and other Kent Coast expresses via both Ashford and Chatham to take the heavier 4-6-0s, the 'King Arthurs' and 'Lord Nelsons'. The Tunbridge Wells-Hastings trains were still in the hands of the various 4-4-0s, but loads were increasing and trains were being accelerated over the London-Tonbridge part of the route with the motive power then available. A locomotive with two outside cylinders large enough to provide the necessary power could not be limited to the clearance required or the axleload limit for the Hastings route. The compromise was to use a shortened version of the 'Lord Nelson' with three cylinders of 16 ½in diameter within the overall width of 8ft 5⁵/16in. A 4-6-0 was not possible with the extreme curvature of the route without even more width restrictions and again existing turntables at Tonbridge and St Leonards were only capable of taking a 4-4-0.

Maunsell therefore designed a 4-4-0 with three cylinders, 6ft 7in coupled wheels and a boiler pressure of 220lbs psi, giving a tractive effort of 25,130lbs at 85 per cent boiler pressure, virtually the same as a 'King Arthur' but less weight overall. Total heating surface was 2,049sqft, including superheating surface of 283sqft. Valve travel was 6½in with lap of 1½in. However, the axleload was at the limit of 21 tons, probably advisable to give more adhesion on so powerful a 4-4-0. Engine weight was 67 tons 2 cwt, tender 42 tons 8 cwt. To conform to the Hastings line gauge, the cab had to be cut back to reduce the width at the top of the kinetic envelope and this required a round-topped rather than Belpaire boiler to give good forward visibility for the driver, and also to reduce weight as the design was already at maximum axleload. In effect, the boiler was a shortened 'King Arthur' boiler, with grate area of 28.3sqft. The result of these compromises was in fact a well-balanced design which was successful from the start and far exceeded the experience with the larger 'Lord Nelsons' which were subject to many experiments in the 1930s to try to improve their steaming and performance. The 'Schools' suffered no such tribulations and although Bulleid later fitted about half the class with wide diameter Lemaître multiple jet exhausts, the change seems to have brought no measurable improvement. The tenders had the capacity for 4,000 gallons of water and five tons of coal. One tender, fitted initially to 932 (and later after that engine's withdrawal, to 30905) had raised sides fitting the cab profile.

The first ten 'Schools' were constructed in 1930, E900 being completed at Eastleigh Works in March of that year (the prefix E as if they were Western Section engines, being built and maintained at Eastleigh even though their intention was for service on the Eastern Section). Each of the initial batch cost £6,092. They were classified as class 'V' and named after public schools, many of which were on Southern Railway territory and did good business with the company. In many cases, the new locomotive was taken to the nearest station to the school for the naming ceremony and pupils from the school invited to examine the exhibited locomotive. Because of the 21 ton axleload the SR Chief Civil Engineer had to bring the Tonbridge-St Leonards infrastructure up to standard and this was not completed until 1931, so the new engines started work mainly on the Dover and Folkestone route via Ashford. Some were even used initially on the South Western Section to Salisbury and Bournemouth, anticipating the move in the later 1930s when more were available. The tendency for a high powered 4-4-0 to be prone to slipping on starting was partly overcome by an easily operated regulator, though most drivers would start cautiously and wait until the train was well on the move before opening right out.

The first engine was named *Eton* after the most famous public school and 901-918 were named after schools in no particular order, although the pupils of those schools might well argue about the size/prestige/founding date influencing the priority, a somewhat provocative subject. *Eton* was named at Waterloo on 26 March 1930 and exhibited at Windsor & Eton station a couple of days afterwards. 901 was exhibited at Winchester in May, 902 at Crowthorne (Wellington) also in May, 903 at Farncombe (Charterhouse) and 904 at Lancing in June, 905 at Tonbridge, 906 at Sherborne and 907 at Herne Hill (Dulwich) in July, 908 and 909 at Waterloo (Westminster and St Paul's) in October. E900-909 were built without smoke deflectors which were subsequently fitted to all new 'Schools' from 910 and retrospectively on the first ten. 919-930 were then named after the most famous schools north of the Thames, the last nine reverting to Southern territory.

The first ten moved to St Leonards for the Charing Cross-Hastings services for which they were designed after the track improvements in 1931 and 910-924 were built from August 1932 to the end of 1933, cost per locomotive coming down to £5,374. Most of these were similarly exhibited and seen by boys from the school although those with names outside SR territory were not. 923 was named *Uppingham* until the Headmaster objected strongly (despite one of the school's trustees being a SR director!). Apparently, the man had strong views against self-advertisement and the nameplates were replaced by ones commemorating Bradfield school, although these were not fixed until August 1934. Despite the furore, Uppingham is said to have the original nameplates in the school museum! A further batch, 925-930, was constructed in 1934 and the final nine in 1935, costing £5,256 and £5,209 (the last five) each.

Maunsell retired in 1937 and was replaced by Oliver Bulleid who concentrated mainly on trying to root out the steaming problems of the 'Lord Nelsons' as his first

Official works-grey photo of the prototype, E900 *Eton*, March 1930. Real Photographs/ MLS Collection

E900 *Eton* at Windsor & Eton station on exhibition for inspection by pupils from the school, 28 March 1930. John Scott-Morgan Collection

priority. However, he initiated some experiments with the 'Schools' class, undertaking some tests with 901 to assess cylinder and motion efficiency. On the whole he ignored the 'King Arthur' class which he found too basic for his ingenuity, although he half-heartedly equipped a few of the Urie engines with multiple-jet Lemaître exhausts with little impact on their performance (apart from 755 which had a few other variations which stood that engine out from the crowd). Bulleid, as stated earlier, then fitted eighteen 'Schools' with multiple-jet exhausts, starting with 914 in

E904 *Lancing* on exhibition at Lancing station while boys from the college swarm all over it, June 1930. John Scott-Morgan Collection

E900 brand new and recently named *Eton* at Charing Cross, March 1930. MLS Collection

903 *Charterhouse* at Cannon Street with a Ramsgate and Dover train during the first few months while it was stationed at Deal pending strengthening of the Tunbridge Wells-Hastings section, and after the fitting of smoke deflectors, 1931. Real Photographs/J.M. Bentley Collection

January 1939 after improvement had been effected with the 'Lord Nelsons' but no further significant modifications were made throughout the thirty years of the locomotives' lives. Bulleid did carry out some cylinder experiments on 937 in May 1939 – possibly resulting from data collected on the tests with 901 – which also involved setting forward the blast pipe and adapting the chimney angle. These may have been some experiments looking forward to his new pacific designs, but the war stopped any development on the 'Schools' and 937, despite its excellent

910 *Merchant Taylors* during its construction at Eastleigh Works, 25 September 1932. It was completed and delivered for traffic in December 1932. J.M. Bentley Collection

916 *Whitgift* shortly after construction, 1933. J.M. Bentley Collection

923, named *Uppingham* when constructed in December 1933 and before renaming in June 1934. J.M. Bentley Collection

930 *Radley* when brand new at Bricklayers Arms depot, alongside one of the shed's 'King Arthurs', January 1935. J.M. Bentley Collection

936 *Cranleigh* nearing completion at Eastleigh Works, 30 July 1935. H.C. Casserley/J.M. Bentley Collection

performance as adapted, reverted to a standard engine with multiple-jet blast arrangements in March 1941.

The SR lined green livery was replaced with a brighter green in the Bulleid era, later termed malachite. Sage-green was also tried on some locomotives including 934 in 1939. In the 1930s, streamlining was in the vogue north of the Thames and Bulleid made a half-hearted attempt to streamline a 'Schools' in March 1938, fitting 935 with a plywood frame streamlined outline and renumbering it 999, but the result frankly looked hideous (at least in my opinion) and thankfully the experiment was abortive and the streamlining was removed in April, the engine returning to its previous identity.

During the Second World War, they operated mainly in the forefront of operations nearest the German lines, so they were vulnerable to enemy attack. (In fact 58 attacks on SR trains, mainly in Kent, were recorded between July 1940 and August 1943 – a couple of drivers were killed in attacks, both Ramsgate men.) One locomotive (931) was fitted with protective armour cladding round the cab, but

901 ***Winchester*** at Cannon Street during tests instituted by Bulleid to check cylinder efficiency, with indicator shelter fitted for protection of engineers whilst the train is in motion, July 1938. Real Photographs/MLS Collection

937 ***Epsom*** at Eastleigh experimentally fitted with revised cylinder and blast pipe arrangements and extended smokebox by Bulleid, 1939. B.W. Anwell/John Scott-Morgan Collection

932 *Blundells* with its unique high-sided tender, at Nine Elms, c1937. J.M. Bentley Collection

the plan to fit 200 engines in South East England was abandoned – apparently crews complained of the heat and claustrophobia, so the experiment was unique. Wartime black replaced malachite and sage green liveries from 1942.

Because of their vulnerability to attack in their sphere of operation between 1939 and 1945, several locomotives received war damage. The most serious incident involved 934, which took a direct hit on 11 May 1942 on Cannon Street Bridge just outside the station but was repaired and re-entered service. 900 was damaged in an incident at North Kent East Junction in

Bournemouth's 925 *Cheltenham* at Eastleigh in the company of an Adams 4-4-0, an 'L12' and two 'King Arthurs', 19 March 1939. W. Potter/ J.M. Bentley Collection

935 *Sevenoaks* covered with experimental plywood streamlining and renumbered 999, March 1938. The second photo is a Southern Railway official photo in its 'Sunny South Sam' guise. John Scott-Morgan Collection

The protective armour surrounding the cab of 931 *King's Wimbledon*, 1943. SR official photograph/John Scott-Morgan Collection

September 1940, 912 was damaged when the roof of Ewer Street depot was hit by a bomb on 10 October 1940 and 936 fell into a bomb crater four days later just south of London Bridge. A South Western Section engine, 927, was caught by the blast when Nine Elms was hit by bombs in April 1941, and required a boiler replacement, 914 was derailed after track damage at New Cross in the same month. 917 was damaged at Deal in August 1942 when one of the Ramsgate drivers was killed. 922 was attacked while standing at Westenhanger station and the other Ramsgate driver was killed and a train hauled by 912 was attacked at the same time and a soldier on the train died. 901 and 904 were slightly damaged in a bomb explosion at Hastings in March 1943. Bricklayers Arms, Stewarts Lane and New Cross Gate depots were all bombed at one stage of the war.

In January 1946, the task of restoring the 'Schools' to pre-war malachite green began with 934, 903, 907 and 917. At nationalisation, the 'Schools' were renumbered

905 *Tonbridge* in wartime black livery at Bricklayers Arms, 22 September 1947. J.D. Darby/MLS Collection

Multiple-jet chimneyed s934 *St Lawrence* still in malachite green SR livery but lettered 'British Railways', at Victoria, 1948. MLS Collection

30917 *Ardingly* newly painted in malachite green but BR number and lettering, ex-works Eastleigh, 5 June 1948. 30917 was the last 'Schools' to retain this livery which it lost in 1952. J.M. Bentley Collection

30934 *St Lawrence* at Eastleigh newly painted in malachite green with BR number and lettering, 11 May 1948. H.C. Casserley/J.M. Bentley Collection

30900-30939, and were surprisingly classified as mixed traffic rather than passenger engines and received the lined black livery, starting with 30923 in October 1948. The last 'Schools' in malachite green was 30917 which received the BR mixed traffic livery in July 1952. The passenger livery of BR Brunswick green was finally applied from January 1957 (at the same time as Swindon Works started painted all their engines green bar the obviously purely freight engines).

With the electrification of the LC&DR route via Chatham in 1959, many 'Schools' lost some of their most important Kent Coast work, but it was mainly the earlier Wainwright/Maunsell rebuilt engines that took the brunt of the withdrawals and the first 'Schools' were not withdrawn until January 1961 (30919 and 30932, the latter with the high-sided tender). Two 'Schools' (30912 and 30921) received eight wheel high sided tenders from withdrawn 'Lord Nelsons' in November 1961. Steady withdrawals took place throughout

30936 *Cranleigh* in mixed traffic lined black and tender without the 'lion & wheel' totem with the rare 0-8-0T 949 *Hecate* in the background, c1950. John Scott-Morgan Collection

30901 *Winchester* in BR mixed traffic lined black livery at St Leonards, 23 June 1951. L. Elsey/ J.M. Bentley Collection

30903 ***Charterhouse*** at its home depot of St Leonards along with a couple of 'R' 0-6-0Ts, 14 July 1954. MLS Collection

Ramsgate's 30911 *Dover* ex-works in the BR lined black livery, c1954.
John Scott-Morgan Collection

30913 ***Christ's Hospital*** ex-works at Eastleigh in the BR mixed traffic livery, 15 May 1954.
John Scott-Morgan Collection

30929 *Malvern* awaiting heavy repair at Ashford, alongside ex-LMS Brighton-built 2-6-4T 42077 and E2 0-6-0T 32100, June 1958. Photomatic/John Scott-Morgan Collection

30935 ex-works in BR passenger Brunswick green and late icon, c1959. MLS Collection

Large diameter chimney 'Schools' 30939 *Leatherhead* at Stewarts Lane, in BR green livery and early small 'lion & wheel' icon, c1958. MLS Collection

30905 *Tonbridge* with high-sided tender off 30932 at Basingstoke, summer 1959. J.M. Bentley Collection

30903 *Charterhouse* of Guildford shed at Nine Elms minus nameplates at its time of withdrawal (with all the remaining 'Schools') December 1962. J.M. Bentley

30914 *Eastbourne* in BR lined black mixed traffic livery at Stewarts Lane, 24 May 1958. R.C. Riley

30901 *Winchester* EX-WORKS in BR Brunswick green passenger livery at Ashford, 20 June 1960. R.C. Riley

30918 *Hurstpierpoint* at Willesden shed after having worked a returning holiday train from the South Coast to the LMR, 10 July 1960. The BR ATC box is very visible on the running plate. R.C. Riley

30921 *Shrewsbury* fitted with 'Lord Nelson' tender off the withdrawn 30854, shunting at Basingstoke, 1961. Ken Wightman

1961 and 1962, but seventeen were still operational at the end of 1962, when the whole class, irrespective of condition, was withdrawn at the stroke of the accountant's pen. Mileage accumulated for most of the class was around a million. Three were preserved, one – 925 *Cheltenham* – under the BR official preservation list, and two under private ownership, 926 *Repton* going initially to the United States, the other, 928 *Stowe*, being purchased for the Duke of Montagu's estate at Beaulieu.

Operation

Although the 'Schools' were designed for the Hastings route in particular and the first engines built were intended for St Leonards depot, the track between Tunbridge Wells and St Leonards still needed bringing up to the standard that could take 21 ton axleload locomotives. This was not completed until 1931, so the first ten locomotives of the class built in 1930 were found initially alternative homes. E900 and E901 were tested in the first few weeks from Eastleigh on the Bournemouth line to ensure no immediate snags, and when all was shown to be well, E900-903, 905/6 went to Deal and worked Charing Cross-Folkestone 80-minute expresses taking over from the 'Ls' and 'L1s' whilst the others were based at Eastbourne. Deal shed closed in September 1930 and the 'Schools' fleet based there were transferred to Ramsgate.

The Tunbridge Wells-St Leonards track condition was ready for the heavier locomotives by July 1931 and on the 6th the 10.25am Charing

E900 at the head of a Folkestone-Charing Cross 80-minute express including Pullman car, 1930. MLS Collection

E900 ***Eton*** on an express for Dover leaving the Shakespeare Tunnel at Folkestone Warren, June 1931. J.M. Bentley Collection

E901 *Winchester* at the head of the 4.15pm Charing Cross *Minster Pullman* express, near Orpington Junction, 28 June 30. K. Nunn/LCGB/ MLS Collection

Eastbourne's E904 *Lancing* approaches Clapham Junction with an express including Pullman cars for Haywards Heath and Eastbourne, 1930. Colling Turner/John Scott-Morgan Collection

E904 ***Lancing*** at Honor Oak Park with an Eastbourne-London Bridge express, just a month before the engine's transfer to St Leonards in July 1931. Photomatic/J.M. Bentley Collection

902 ***Wellington*** on a Hastings express near Mountfield Tunnel, c1932. J.M. Bentley Collection

923, still named *Uppingham,* on a Charing Cross-Hastings express departing from Wadhurst station, 1934. It received its new *Bradfield* nameplates in August of that year. J.M. Bentley Collection

900 *Eton* at the head of a Hastings express at Cannon Street alongside 'N15' 800 *Sir Meleaus de Lile* on a Dover train, c1933. John Scott-Morgan Collection

900 ***Eton*** at Mountfield Halt near Battle on a Charing Cross-Hastings express, c1934. J.M. Bentley Collection

Cross with E904 took a 390 ton load including special vehicles for the Southern Railway management to open the new Hastings station and experience the advent of the new power for the line. The crew and engine regained five minutes of the ten minutes late start and E909 returned the party home on the 4.55pm. St Leonards depot was then allocated four of the initial ten, the other six being at Ramsgate, Eastbourne losing 'Schools' which were replaced by the 3-cylinder 'U1' moguls.

A second batch of 'Schools' came on stream in early 1933 and 911 and 912 (the 'E' was dropped at the end of 1931) were allocated to Ramsgate and 913-916 were back to Eastbourne supplanting the 'U1s' and restoring the high performance of the 'Schools' on that route for the eighteen months before full electrification in July 1935. The summer 1933 allocation was:

Ramsgate: 900-902, 910-912, 919
St Leonards: 903-909, 917, 918
Eastbourne: 913-916

A number of locomotive performance recorders were active in checking out the powerful new 4-4-0s especially on the Charing Cross-Folkestone and Hastings expresses and to a lesser extent on the Eastbourne route and I give a few examples of their early work which were publicised at the time. The runs on the next page were the Charing Cross-Folkestone 80-minute expresses (the runs with 922 below were scheduled 80 minutes including the Waterloo East stop). It will be noted that some recorders timed to the nearest five seconds.

Charing Cross-Folkestone Central

		916 *Whitgift* – Ramsgate 9 chs, 282/300 tons			922 *Marlborough* – Ramsgate 8 chs, 254/275 tons		
Miles	**Location**	**Times**	**Speeds**	**Punctuality**	**Times**	**Speeds**	**Punctuality**
0.0	Charing Cross	00.00		T			
0.8	Waterloo East	02.25			00.00		T
1.9	London Bridge	04.45		½ E	04.05	sigs	2 L
4.9	New Cross	10.09	sigs	¾ L	08.05		
7.2	Hither Green	13.02		½ L	11.00		
11.2	Chislehurst	18.54	38	1 L	16.25		
13.8	Orpington	22.14	49	1 ¼ L	19.40	52	1¾ L
16.6	Knockholt	25.48	45		23.00	50	
20.6	Dunton Green	29.56	72		27.30	65	
22.1	Sevenoaks	31.18		¼ L	29.45	pws	¾ L
27.0	Hildenborough	35.41	78		34.40	76	
29.5	Tonbridge	38.02		½ E	37.25		
34.8	Paddock Wood	43.15		1¼ E	43.15		
39.4	Marden	47.14			47.00		
41.9	Staplehurst	49.28	71	1½ E	49.10	73	
45.3	Headcorn	52.26			52.00		
50.5	Pluckley	57.17			56.45		
56.1	Ashford	62.33		2½ E	61.40		¼ E
60.4	Smeeth	66.48	pws		65.45		
65.4	Sandling Jcn	75.10		¼ L	70.50		
69.3	Shorncliffe	79.02			74.30		
70.0	Folkestone Ctl	80.08	(76 ½ net)	T	76.25	(74 net)	½ E

Ramsgate's 912 *Downside* approaches Orpington with a Charing Cross-Folkestone Central express, c1935. J.M. Bentley Collection

		Folkestone Central-Charing Cross 903 *Charterhouse* – Ramsgate 10 chs, 329/355 tons			922 *Marlborough* – Ramsgate 9 chs, 290/305 tons		
Miles	**Location**	**Times**	**Speeds**	**Punctuality**	**Times**	**Speeds**	**Punctuality**
0.0	Folkestone Ctl	00.00		T	00.00		T
4.5	Sandling Jcn	08.15	50	¼ L	07.22		T
9.6	Smeeth	13.15	75		11.56	81	
13.8	Ashford	16.40	pws	¼ E	15.13		1 ¼ E
19.5	Pluckley	23.40			19.39		
24.7	Headcorn	28.35	72		23.37	80	
28.1	Staplehurst	31.25	78		26.14		
30.6	Marden	33.35	68		28.11	75	
35.1	Paddock Wood	37.20	74	1¼ L	31.48	74/76	2¾ E
40.4	Tonbridge	42.10		1¼ L	36.54		2½ E
42.9	Hildenborough	46.35	32		40.38	39	
47.8	Sevenoaks	54.45		1¾ L	47.45		3¾ E
49.4	Dunton Green	56.25	34		49.21	35	
53.4	Knockholt	60.50	48	1¾ L	53.50	53/64	
56.1	Orpington	63.15		1¼ L	56.00		4 E
58.7	Chislehurst	65.20	78	¾ L	58.12	66	4¼ E
60.9	Grove Park	67.05	78/ 60*		60.14	72	
62.8	Hither Green	68.40		¾ L	61.23		4½ E
65.1	New Cross	71.45		¾ L	64.14		4¾ E
68.1	London Bridge	76.00		1 L	72.17 sigs stand (2 mins)		
69.2	Waterloo East	-	sigs		79.34 sig stop (72 net)		3½ L
70.0	Charing Cross	82.45	(77 ½ net)	2¾ L			

Whilst the run with 903 above was close to schedule, but with a couple of coaches over the normal load, 922 was pushed hard on both outward and inward journeys to keep to the faster schedule which included the Waterloo East stop. The timekeeping on both up journeys was spoiled by signal checks from London Bridge.

The 'Schools' were designed specifically for the Hastings route and so it is of interest to show what was required on a normal Charing Cross-Hastings express. Again, the gradient profile is shown in the

936 *Cranleigh* on the Petts Wood loop with a Continental Boat Train, c1935. P. Rutherford/J.M. Bentley Collection

923 *Bradfield* at Charing Cross with an express for Folkestone and Dover, c1935. J.M. Bentley Collection

appendix, but the key gradients south of Tonbridge are pertinent to the log on the next page. Out of Tonbridge, there is an immediate short rise of 1 in 53/47 followed by five miles continuous rise at around 1 in 80/1 in 100 beyond Tunbridge Wells to Strawberry Hill Tunnel just before Frant. There are then three miles of 1 in 100 down/1 in 118 up to Wadhurst and then eleven miles of falling gradients ranging from 1 in 100 to 1 in 150 but with speed restriction for a couple of miles south of Wadhurst Tunnel. There is then a sharp rise at 1 in 86 to Mountfield Tunnel, easing through Mountfield Halt, then three miles of 1 in 100/132 to Battle, slight fall to Crowhurst, where the recorder alighted.

Charing Cross-Crowhurst
906 *Sherborne* – St Leonards
8 chs, 261/275 tons

Miles	Location	Times	Speeds	Punctuality
0.0	Charing Cross	00.00	sigs	T
1.9	London Bridge	05.15		¼ L
4.9	New Cross	10.00	sigs	1 L
7.2	Hither Green	12.44		¾ L
11.2	Chislehurst	18.47		¾ L
13.8	Orpington	22.12		¾ L
16.6	Knockholt	25.43	48	
20.6	Dunton Green	30.03	65	
22.1	Sevenoaks	31.32	30*	½ L
27.0	Hildenborough	36.23	75	
29.5	Tonbridge	39.19		¼ L
32.9	High Brooms	46.00	31	
34.4	Tunbridge Wells	48.51	32	¼ E
36.7	Frant	52.54		
39.3	Wadhurst	56.28	45*	
43.9	Ticehurst Road	62.11	50	
47.5	Etchingham	65.22	67/75	¾ E
49.6	Robertsbridge	67.24	60	
55.6	Battle	74.53	48/40	¾ E
57.7	Crowhurst	78.13	(77 net)	¾ E

Crowhurst-Tunbridge Wells
908 *Westminster* – St Leonards
11 chs, 357/380 tons

Miles	Location	Times	Speeds	Punctuality
0.0	Crowhurst	00.00		T
2.1	Battle	-	67	
	Mountfield	-	45	
8.1	Robertsbridge	-	67	
10.2	Etchingham	13.51		
13.8	Ticehurst Road	-	47	
18.4	Wadhurst	24.25	36	2½ E
23.3	Tunbridge Wells	31.21		3½ E

This run was made with the maximum load for a 'Schools' on this route.

The 'Schools' only had a brief sojourn on the Eastbourne main line as it was electrified within a couple of years of the class being available in sufficient numbers to cover the key turns on a regular basis. Very few logs are in the public realm, but luckily a few were recorded by H.T. Clements, the main interest being the run as far as Haywards Heath on the Brighton main line before turning off for Lewes and Eastbourne.

Bricklayers Arms' 936 *Cranleigh*, just one month after construction, with a Charing Cross-Hastings express near Etchingham south of Tunbridge Wells, July 1935. J.M. Bentley Collection

Victoria-Haywards Heath

		915 *Brighton* – Eastbourne 7 chs, 220/235 tons 18.10.1933			913 *Christ's Hospital* - Eastbourne 11 chs, 350/370 tons 24.9.1933		
Miles	**Location**	**Times**	**Speeds**	**Punctuality**	**Times**	**Speeds**	**Punctuality**
0.0	Victoria	00.00		T	00.00		T
2.7	Clapham Jcn	07.05	sigs		06.50 pws		
10.6	East Croydon	18.45		1¾ L	19.55		3 L
0.0		00.00			00.00		
0.9	South Croydon	02.40			03.20		
3.0	Purley	05.50	46		06.50	40	
4.5	Coulsdon N	07.55	25*	1¾ L	-	sig stand	
8.3	Quarry s/box	13.50	39		20.20	29	
11.3	Earlswood	16.50		1½ L	24.20		10 ¼ L
15.5	Horley	20.20	73		28.00	71	
19.0	Three Bridges	23.40		½ L	31.25		9½ L
21.4	Balcombe Tnl	26.10	52		34.10	47	
23.6	Balcombe	28.40	64		36.40	67	
27.5	Haywards H	35.30	sigs (33 ¼ net)	2¼ L	41.05	(34 ¼ net)	9L

Haywards Heath-East Croydon

		914 *Eastbourne* - Eastbourne 7 chs, 220/235 tons 5.5.1933			910 *Merchant Taylors* –Stewarts Lane 8 chs, 265/285 tons 26.4.1935		
Miles	**Location**	**Times**	**Speeds**	**Punctuality**	**Times**	**Speeds**	**Punctuality**
0.0	Haywards H	00.00	30* (pass)	¼ E	00.00		T
3.9	Balcombe	04.45	50		06.50	48	
8.5	Three Bridges	09.40	pws 40*	¾ E	11.50	60	¼ E
12.0	Horley	14.15	60		14.55	70	
16.2	Earlswood	18.30	55	¼ L	18.50		1¼ E
19.3	Quarry s/box	22.10	sigs 40*		22.10	50	
23.0	Coulsdon N	-	62		26.55	sigs 30*	1 E
24.5	Purley	28.15	70		29.05	sigs 30*	
26.6	South Croydon	32.40	sigs		32.00		
27.5	East Croydon	34.35	(30 ½ net)	2¼ L	33.40	(31 ½ net)	¼ E

A batch of 'Schools' built in 1934/5 (924 -933) was based at Fratton for work on the Portsmouth direct expresses, prior to electrification in 1937. They replaced 'D15s', 'L12s', 'U1s', 'H15s' and 'S15s' on the fastest trains and were a significant step forward. The allocation was rationalised after the electrification of the Victoria-Eastbourne route, and the distribution from August 1935 was:

St Leonards: 900-911
Ramsgate: 912-923
Fratton: 924-933
Bricklayers Arms: 934-939

Cecil J. Allen in his monthly article of February 1936 in the *Railway Magazine* recorded a selection of the Fratton engines' work on the Portsmouth line and

911 *Dover* at Battle station with a Hastings express, 25 July 1935. J.M. Bentley Collection

I have chosen four of these as representative.

The first run is a good timekeeping performance with the standard load and one permanent way slack. The third and fourth runs are with an extra coach, with 933 on the fourth run performing energetically after the early p-way slack and then easing when timekeeping was assured, whilst 927 on the third run recovered slowly from the p-way restriction and then hurried in the latter part of the journey recording two 80s. The second run with 931 was a 'tour de force'. After sustaining very bad delays in the early stages, 931 was

914 *Eastbourne*, shedded at the home depot, celebrates the 1935 Silver Jubilee of King George V and Queen Mary with special decoration ready for the town's celebrations, May 1935. John Scott-Morgan Collection

Waterloo-Portsmouth 1934/35

	3.50pm Waterloo 924 *Haileybury* 10 chs, 318/340t			6.50pm Waterloo 931 *King's Wimbledon* 10chs, 324/345t			11.50am Waterloo 927 *Clifton* 11chs, 350/380t			Unidentified 933 *King's Canterbury* 11chs, 356/385t		
Location	**Times**	**Speeds**		**Times**	**Speeds**		**Times**	**Speeds**		**Times**	**Speeds**	
Waterloo	00.00			00.00			00.00			00.00		
Clapham Jcn	07.13	pws 25*			sigs stand		07.57	pws 30*		07.49	pws 25*	
Wimbledon	12.30			15.32	pws 20*		12.42			12.47		
Surbiton	18.20			24.13			18.23			17.54		
Hampton C Jcn	19.38	65	2 L	26.11	56	8¾ L	19.40	65	2 ¼ L	19.08	67	1¾ L
Weybridge	23.08	60/69		30.10	67/69		23.07	58/66½		22.29	63/68	
West Byfleet	27.25			34.28			27.29			26.38		
Woking	30.09	35*	1¾ L	36.55	38*	8½ L	30.03	30*	1½ L	29.16	35*	½ L
Worplesdon	33.21	61		39.50	63		33.46	59		32.35	61½	
Guildford	37.50	30*	1¾ L	43.19	35*	7¼ L	38.01	30*	2 L	36.52	28*	¾ L
Godalming	43.05	58		47.48	65		44.03	55		42.32	58	
Milford	45.04			49.30			46.07			44.10		
Whitley	47.50	41/58		51.48	54/62		49.00	40/56		47.44	34½ /55	
Haslemere	55.23	27½	¾ L	57.20	36	2¾ L	56.41	21	2¼ L	56.19	22	1¾ L
Liphook	59.18	79		60.42	79		60.29	82		60.15	71½	
Liss	63.17			64.26	65*		64.21			64.22		
Petersfield	66.12		¾ E	67.10	73	¼ L	67.08		¼ L	68.11		1¼ L
Buriton Sidings	68.42	56		69.20	54/74		69.32	42/80		71.21	37½ /67½	
Rolands Castle	75.36	49*		75.04	51*		75.28	52*		78.18		
Havant	79.09			78.49	pws 22*/66		79.00	sig stand (44s)		81.42	63	
Fratton	87.12	(85 net)	1¾ E	86.43	(78 net)	2¼ E	88.52	(86 net)	T	88.37		½ L
				00.00								
Portsmouth				02.52						90.33	(88½ net)	½ L

driven hard after Woking and the uphill work was exceptional. O.S. Nock had a couple of footplate runs on these services around this time, with 924 *Haileybury* on the 11.50am Waterloo with a 350 ton load and 925 *Cheltenham* on a packed 10.10am Sunday morning up service, loading 395 tons gross. The down run was very similar to runs in the table above, Havant being passed on time (81 minutes for the 66.4 miles inclusive of two severe p-way slacks and a signal check before Guildford), but the line was badly congested on to Portsmouth Harbour with holiday traffic for the Isle of Wight and arrival was 24 minutes late. 925 on Sunday was superb despite the overload – highlights were 45/25mph on the 1 in 110/1 in 80 climb to Buriton summit, passed 4½ minutes late after three signal checks before Havant, 75/83mph before and after Petersfield, 48½ minimum at Haslemere at the summit of the 1 in 100 (now just 1½ minutes late) and 74 before a p-way slack at Witley. Another 15mph p-way slowing at Shalford Junction caused the train to be 2¼ minutes late past Guildford but sustained running at 75-78mph between Walton and Hampton Court Junction had recouped a minute before another p-way slack at New Malden lost it again. Clapham Junction was passed just two minutes late but a dead stand at Queen's Road meant a four minute late arrival (net time 82 minutes against the scheduled 90).

The Rev R.S. Haines kept a very detailed log of his frequent

Two Fratton 'Schools' on shed shortly after their allocation there, 931 *King's Wimbledon* and 932 *Blundells* before the latter acquired a high-sided tender, 26 October 1935. H.C. Casserley/J.M. Bentley Collection

journeys between Fratton, Havant and Petersfield between 1935 and 1937 and recorded an extraordinary variety of classes on the local and semi-fast services he used, details of which are now kept for posterity in the archives of the Railway Performance Society. He kept outline times for each run and analysed them as exceptional, very good, good, average or poor. He obviously knew the drivers well and experienced some (unofficial) footplate rides. Summarising a 13-week period after the summer of 1936, he said:

> '… the standard the locomotive performance reached during these weeks has been good. This is taken as a whole, but it might have been far better if runs behind the (ex-LBSCR) 4-4-0 'B4Xs' had been omitted…. they make a good deal of noise caused by escaping steam. The Brighton tanks ('I3') do fairly well, chiefly due to their rapid acceleration. On the Portsmouth direct line a very much higher standard is reached. This is in some measure due to the excellent 'Schools' engines which nearly always display a really good performance. Some of this is due to the drivers who work them. These drivers get the utmost out of them when running behind time, but no thrashing is required on an unchecked run from London-Portsmouth on the 90-minute schedule. In fact the 'Schools' would be capable of covering the 72 miles in 80 minutes if the load were limited to seven coaches and a little thrashing now and again.'

He goes on to comment that the 'King Arthurs', normally driven by Nine Elms men, do not often show the same vigour although he ascribes this to the attitude of the men rather than the

A Fratton 'Schools', 925 *Cheltenham*, with a Waterloo-Portsmouth express, c1935. John Scott-Morgan Collection

926 *Repton* passing Queen's Road Battersea with a Waterloo-Portsmouth express, 16 May 1936. H.F. Wheeler/ John Scott-Morgan Collection

engines and he has noted several exceptional runs with both the Urie and 'Scotch Arthurs', but remarks on the poor performance of the 3-cylinder 'U1' moguls and the 'T14' 4-6-0 'Paddleboxes' especially uphill.

Rev. Haines' summary of the working in the Autumn of 1936 further endorses the performance of the 'Schools'. I quote:

> 'The 'Schools' engines continue to provide excellent performance. Whatever may be their drivers, they seldom,

933 ***King's Canterbury*** entering Godalming station with a Waterloo-Portsmouth train, 29 February 1936. H.C. Casserley/J.M. Bentley Collection

The Royal Train composed of six Pullman and 915 *Brighton* cars and luggage van at Herne Hill en route from Victoria to Dover on the State visit of King George VI and Queen Elizabeth to Paris, 19 July 1938. John Scott-Morgan Collection

very seldom in fact give a poor run. Whether it is with a 4-coach train or 11 coaches, they always do their job well. They seem easy to manage, they are economical in coal consumption and make little noise when running. Sentimentally, they are also handsome to look at.'

After the electrification of the direct line to Portsmouth in 1937, the ten Fratton 'Schools' were transferred to Bournemouth replacing ten 'Scotch Arthurs' that had been doing valiant work over the previous ten years. Despite their lower adhesion, the work of these 4-4-0s prior to the advent of the Second World War was often brilliant and I record on the next page the logs of a number of runs from this time, some on the non-stop *Bournemouth Limited*, allowed 116 minutes for the 108 miles. The first group of four runs are all with loads substantially greater than set for the schedules in operation at the time.

The Royal Train passing Bromley South, 19 July 1938. J.M. Bentley Collection

Waterloo-Bournemouth, 1937-9

	932 *Blundells* 12 chs 386/415t *Bournemouth* Ltd			926 *Repton* 14 chs 448/490t 6.30pm Waterloo			927 *Clifton* 15 chs 491/525t Driver Allen (B'mouth)			932 *Blundells* 15 chs 473/510t Driver Allen		
Location	**mins secs**	**mph**		**mins secs**	**mph**		**mins secs**	**mph**		**mins secs**	**mph**	
Waterloo	00.00			00.00			00.00			00.00		
Clapham Jn	07.53		¾ L	07.41		¾ L	08.00		1 L	08.13		1¼ L
Wimbledon	12.09			12.32			12.27			13.07	47	
Surbiton	17.21	61 ½		18.09	60		17.53	62		18.40	58	
Weybridge	24.30	56		25.32			25.01	56		25.24	60/67	
Woking	29.40	65	1¾ L	31.16	61 ½	1¾ L	30.16	63½	¾ L	30.22	59	¾ L
Brookwood	33.36			35.40			34.30			34.18	53	
MP 31	37.04	51		39.33	45		38.18	46½		37.52	48½	
Fleet	42.37			45.41			44.15			43.34	65½	
Hook	47.57	66 ½		51.37	61		50.01	57/63½		49.03	59/65½	
Basingstoke	53.06			57.19		3¼ L	55.34		1½ L	54.20		¼ L
Worting Jn	55.40	55	1¾ L	60.18		3¼ L	58.27	45	1½ L	57.08	50	½ T
Wootton	58.13	52		63.14	47		61.20			59.55	47½	
Micheldever	63.40	73		69.27			67.33	76		65.52	64	
Winchester	70.08	86		76.17	86 ½	2¾ L	73.36	sigs sl.	T	72.34	82	½ E
Eastleigh	75.07	62*	½ E	81.02	90	L	80.11	66	¾ L	77.52	60*	1½ E
St Denys	78.35			84.02	sig stand		83.34			81.26	62	
Northam Jn	80.11	20*	¼ E	88.32		4 L	85.00	22*	½ L	83.20	15*	1 E
Southampton Central	82.34		½ E	91.16	(88 net)	3¾ L	88.00	(87 ½ net)	½ L	86.32		1 E
				00.00						0000		
Lyndhurst Rd	91.01	47		10.49	43							
Beaulieu Rd	-	56½/			53							
Brockenhurst	98.27	69		18.31	64							
Sway	101.18	55		21.30	53							
Hinton Admiral	106.24	76		26.52	79							
Christchurch	109.26	44*		29.39	47*					30.06 sigs		
Bournemouth	115.52	(114½ net)	T	35.44		¼ E				35.45		¼ E

The *Bournemouth Limited* was one coach over the stipulated load and 932's fireman was struggling with poor coal causing bad steaming in the early part of the run. The other runs, with extremely heavy loads, lost time to Basingstoke and then in typical 'Schools' fashion romped away downhill. 932 with Driver Allen of Bournemouth again did particularly well with the 15 coach 6.30pm Waterloo and having regained time by Worting Junction, was not pushed so hard downhill.

In the up direction, the non-stop *Bournemouth Limited* was allowed 118 minutes and I show on the next page three runs with this train, the first two with the booked 10 coach formation and the third with two extra coaches.

Bournemouth – Waterloo (The *Bournemouth Limited*) 1938/9

	930 *Radley* **10 chs 323/345t**			**931 *King's Wimbledon*** **10 chs 323/345t**			**929 *Malvern*** **12 chs 386/415t**		
Location	**mins secs**	**speed**		**mins secs**	**speed**		**mins secs**	**speed**	
Bournemouth	00.00			00.00			00.00		
Christhurch	06.25	64		06.38	63		06.59		
Hinton Admiral	09.56	53		10.16	52		11.00	45	
Sway	15.40			18.09	pws		17.26	pws	
Brockenhurst	19.17	pws	¼ L	20.46		1¾ L	21.41	64	2¾ L
Beaulieu Rd	23.33	70		24.44	75		26.16	59/67	
Lyndhurst Rd	25.54	64		27.15	pws		28.52		
Redbridge	29.18			36.13	pws		32.15	pws	
Southampton Central				32.32			39.31	38.28	
Northam Jn	34.26		½ E	41.32	18*	6½ L	40.55		6 L
Eastleigh	40.33	54½	½ E	47.52	59	6¾ L	47.49	54	6¾ L
Shawford	44.57	53		51.50			52.17		
Winchester	48.23	58		55.14			55.06	46	
Wallers Ash	53.27			60.41	53		61.06		
Micheldever	57.30	54½		64.41	55		66.35		
Litchfield	59.22			66.37			68.56		
Worting Jn	65.36		4 E	72.29		3 L	75.16		6¼ L
Basingstoke	67.59			74.35			77.32		
Hook	72.37	73½		78.53	80		82.04	75 ½	
Fleet	77.26	69/75		83.20	74½ / 78½		86.44	74	
MP 31	82.20	62		87.46	71		91.16	70	
Woking	88.18	70½	3¾ E	96.02	pws 20*	4 L	96.28	81	4½ L
Weybridge	92.58			100.18	78/72		100.24	80/75	
Surbiton	99.51	pws 18*	3¼ E	106.05	78½	3 L	106.10	70/pws	3L
Wimbledon	106.41			110.15	67½/72		113.16		
Clapham Jn	110.48		¼ E	113.20	35*	2¼ L	116.45		5¾ L
Vauxhall	114.20	sigs 10*		116.57			120.00		
Waterloo	118.05	(113 ½ net)	T	119.26	(108 net)	1½ L	122.24	(114½ net)	4½ L

Cecil J. Allen published a further selection of runs on the Bournemouth road with 'Schools' in the *Railway Magazine* just before the Second World War. 925 *Cheltenham* on the non-stop *Bournemouth Limited* with a similar load to the run in the table above, arrived over four minutes early. As far as Worting Junction, it was not superior to a 'Scotch Arthur' (CJA's judgement) with top speeds of 68mph at Hampton Court Junction and Hook, but thereafter it flew sustaining 83-86mph from Winchester Junction until brakes went on for the Eastleigh slack, passing Southampton Central in 80 minutes exactly. 929 *Malvern* made a similar overall time with higher speeds earlier – 71 at Hampton Court Junction, 73 at West Weybridge, 82 at Winchester Junction and 75 down Hinton Admiral bank. He also

932 *Blundells,* after provision with the unique high-sided tender, in the cutting just west of Woking Junction with the down *Bournemouth Limited,* 4.30pm Waterloo, passing an Adams '0398' 0-6-0 goods engine on the down slow line, c 1938. F. Foote/John Scott-Morgan Collection

925 *Cheltenham,* now allocated to Bournemouth depot, speeds past Brookwood with an up Bournemouth-Waterloo express, c1938. J.M. Bentley Collection

933 *King's Canterbury* with the Birkenhead-Bournemouth cross-country train which the 'Schools' has worked from Oxford, at Southcote Junction near Reading, c 1938. The GW initially banned the use of SR 4-4-0 locomotives on these heavy trains over its metals and Bournemouth retained two 'N15s', 783 and 784, for these duties, but by this time the GW had relented when the reputation of the Bournemouth 'Schools' had reached those responsible. M.W. Earley/ John Scott-Morgan

927 *Clifton* departs Waterloo with an express for Bournemouth & Weymouth, 24 August 1938. H.C. Casserley/J.M. Bentley Collection

published two runs with heavier loads on the 12.30pm Waterloo with its 87½ minute schedule to the Southampton Central stop. 924 *Haileybury* with 12 coaches, 410 tons gross, arrived a minute early and 925 (again) with 14 coaches, 485 tons gross, arrived on time despite a signal check to 20mph at Winchester Junction, after which it accelerated hard to achieve 82mph after Shawford.

The 'Schools' worked mainly on the Bournemouth line on the South Western section, although occasionally one would break the monopoly of 'King Arthurs' on the West of England road; they rarely ventured west of Salisbury. One run recorded from Waterloo to Salisbury in 1939 with 931 fitted with the multiple jet exhaust was exceptional – driven by the redoubtable Driver Silk of Nine Elms. The train was a relief to the West of England, possibly to the *Atlantic Coast Express*. Unfortunately, the timings were only recorded to the nearest five seconds.

I now select the records of some runs by 'Schools' over their traditional route, published in Cecil J. Allen's regular feature in the *Railway Magazine* (September 1940), but timed and submitted by a regular traveller on the route, A.J. Baker. I have deliberately chosen runs in the up direction, which is slightly tougher than the down, particularly involving the five miles of 1 in 100 from just after Rochester Bridge Junction to Sole Street. Other significant gradients include 1 in 132/101 out of Margate, 1 in 100 to Herne Bay after the dip beyond Birchington, 1 in 132 after Faversham, and 1 in 110 between Teynham and Sittingbourne.

The 'Schools' with modified cylinders, revised blast pipe arrangements and extended smokebox, 937 *Epsom*, having arrived at Waterloo and waiting for release to Nine Elms alongside Urie 'Arthur' 738 *King Pellinore*, 1939. Real Photographs/J.M. Bentley Collection

Waterloo-Salisbury, 1939
****931 *King's Wimbledon***
9 chs, 288/305t

Location	min secs	mph	
Waterloo	00.00		
Vauxhall	03.40		
Clapham Jn	07.10	42½	¼ L
Earlsfield	09.05	53	
Wimbledon	10.55	58	
New Malden	13.25	60	
Surbiton	15.20	69	
Esher	17.15	75	
Walton	19.30	72	
Weybridge	21.15	68½ /75	
Woking	25.30	74½	4½ E
MP 31	-	72	
Farnborough	32.45	73½	
Fleet	35.15	82	
Winchfield	37.45	79½	
Hook	39.35	78½	
Basingstoke	43.50	79	9¼ E
Worting Jn	-	63	
Oakley	48.00	66	
Overton	50.45	70	
Whitchurch	53.40	74	
Hurstbourne	55.00	86/90	
Andover Jn	58.45	85/88	
Grateley	64.00	73	
Porton	71.20/74.30 sig stand		
Salisbury	81.50	(76 net)	6¼ E

5.33pm Margate-Victoria, 1938/9

		921 *Shrewsbury* 7 chs 232/245t Driver Ovenden			920 *Rugby* 8 chs 279/295t			912 *Downside* 9 chs 296/315t			**914 *Eastbourne* 10chs 323/350t Driver Ovenden		
Miles	**Location**	**mins secs**	**mph**		**mins secs**	**mph**		**mins secs**	**mph**		**mins secs**	**mph**	
0.0	Margate	00.00			00.00			00.00			00.00		
3.2	Birchington	07.26	pws/67		06.09	64		06.21	68		05.34	65	
11.2	Herne Bay	15.25	45		14.17	46		14.13	48		13.52	42	
14.8	Whitstable	18.49	71/45*		17.31	70		17.30			17.21	68	
	Graveney Sdg	24.09	sigs		21.55	65		22.00	67		21.40	63	
21.9	Faversham	28.46	60*	3¾ L	24.59	60*	T	25.14	40*	¼ L	24.43	60*	¼ E
25.9	Teynham	33.06	74/53½		29.47	52		30.06	68/52		29.21	51	
29.2	Sittingbourne	35.55	68	2½ L	32.41	65	¼ L	33.04	66	½ L	32.18		¼ E
32.3	Newington	39.01	71½		35.51	sigs stand		36.16			35.49	sigs	
38.0	Gillingham	43.59			47.08			41.31			44.23		
39.6	Chatham	45.49		1¾ L	49.17	sigs	5¼ L	43.30	40*	½ E	47.21		3¼ L
40.9	Rochester Bdge Jn	48.11			52.03			45.38	35*		49.30	34*	
43.0	Cuxton Rd	51.37	35		55.29	39		49.37			52.57	40	
47.0	Sole Street	57.59	40	1 L	62.16	33		58.12	24		60.12	31	
50.5	Fawkham	61.39			66.15			62.30			63.59		
53.4	Farningham Rd	63.50	84		68.34	81		64.44	83		66.14	83	
56.2	Swanley Jn	69.31	pws	2½ L	71.20	54	4¼ L	67.11		¼ L	69.46	pws	2¾ L
59.1	St Mary Cray	73.34			73.57			70.01			73.33		
61.3	Bickley Jn	75.50		2¾ L	76.03		3 L	72.46	sigs	¼ E	75.56	61	3 L
63.0	Bromley South	77.35			77.44	73		74.46			77.41	48/67	
65.2	Beckenham	80.20		2¼ L	80.03	45*	2 L	78.02		T	80.21	42*	¼ L
68.2	Sydenham Hill	84.13			84.02	54/41		81.45	sigs		83.55	54/44	
69.9	Herne Hill	86.29		1½ L	86.07	57	1 L	84.18		¾ E	86.01	57	1 L
73.3	Grosvenor Rd	92.02			91.20			89.33			90.50		
73.9	Victoria	93.41	(83 ¾ net)	1¾ L	93.36	(86 ¾ net)	1½ L	91.26	(90 ¼ net)	½ E	92.48	(86 net)	¾ L

** 914 fitted Lemaitre exhaust. All the others were still fitted with their original Maunsell chimneys, although 920 and 921 were later also equipped by Bulleid with the Lemaitre exhaust system.

Other up runs recorded in the *Railway Magazine* at this time included the following highlights:

Loco/Load	**922 7, 232/245t**	**921 9, 292/315t**	**916 9, 296/315t**	**917 9, 296/320t**	**919 11, 352/380t**
Sole St (min)	30	33	26	sigs -	26
Farningham Rd (max)	80	78	81½	84	81
Net time	87½	88½	89¼	87½	91¾

The cylinder and smokebox modifications made by Bulleid to 937 *Epsom* in 1939 caused interest in the performance of this locomotive and tests were carried out on the 10.35am Waterloo-West of England as far as Salisbury, the return journey being conducted by driver Rice of Nine Elms with a technical observer on the footplate.

Waterloo-Salisbury & return, 12.7.1939
937 *Epsom* – Nine Elms (for test purposes)
12 chs, 388/410 tons
Driver Rice – Nine Elms

Miles	Location	Times	Speeds	Punctuality
0.0	Waterloo	00.00		T
3.9	Clapham Junction	06.45		
7.2	Wimbledon	11.15	sigs	
9.8	New Malden	14.35	sigs	
12.0	Surbiton	16.52	62	
13.3	Hampton Crt Junction	18.08		1¼ L
19.1	Weybridge	23.11	72	
24.3	Woking	28.19	sigs 24*	¼ L
28.0	Brookwood	34.16		
31.0	MP 31	37.33	59	
36.5	Fleet	42.35	69	
39.8	Winchfield	45.26	67	
42.2	Hook	47.32	69	
47.8	Basingstoke	52.14	68	
50.3	Worting Junction	55.01	52	1 L
52.4	Oakley	57.21	60	
55.6	Overton	60.24	68 /sigs	
59.2	Whitchurch	63.29	58	
61.1	Hurstbourne	65.17	70	
66.4	Andover Junction	69.25	84	
72.7	Grateley	74.32	60	
78.2	Porton	79.20	85	
82.6	Tunnel Junction	82.42	15*	¾ E
83.7	Salisbury	84.53	(80 net)	1¼ E

The return journey was completed in 80 minutes 55 seconds, five minutes under schedule, including a dead stand at Woking. The load was the same as on the down journey and top speeds were 84mph at Andover Junction, 69 minimum at Overton, 84 at Hook and 79 at Brookwood before a special stop at Woking to set down the CME party. Woking to Waterloo start to stop was accomplished in 25 minutes 36 seconds, an astonishing time with this load, with 69 at Weybridge, 78 at Hampton Court Junction and still 75 at Wimbledon (there was a 60mph speed limit in later years from New Malden inwards). There is no indication of the speed round Clapham curve and a time of 5 minutes 29 seconds in from there to a stand at Waterloo suggests some speeds over the permitted level. Net time for the 83.7 miles was calculated at 76¼ minutes. The conversion of 937 must have been considered a success, but no more were modified (in fact 937 was converted back to standard form at its next Works visit). The war curbed further development and Bulleid's interest by this time was mainly in the development of his pacifics.

During the six years of the Second World War, the 'Schools' were in the line of fire from enemy aircraft and a number were damaged from strafing attacks, derailments after track damage, or from bomb attacks as described earlier in this chapter. Train timings were slowed as loads increased and track repairs were put on hold except for repairs after damage. Haulage of troop trains, evacuation trains and rescue of the army from Dunkirk figured heavily in their duties. The 'Lord Nelsons' were all congregated on the Western Section, so shared Bournemouth turns with its 'Schools' and some of the Ramsgate and St Leonards engines moved to Stewarts Lane away from the threat to the Kent Coast depots, although South-East London was hardly safe. With the influx of the larger 4-6-0s, some of the Bournemouth 'Schools' were moved to Basingstoke in 1943 although it was not until 1945 that the majority moved away from the Western Section back to St Leonards and Brighton. The Brighton engines worked to Bournemouth and to Salisbury on Plymouth and Cardiff services. In the immediate aftermath of the war, the Southern Railway announced plans to electrify all the remaining lines in the Eastern Section including the Hastings route. However, these plans were shelved with the changes of management incurred at nationalisation and it was to be another ten years before the 'Schools' lost their main workings there – although they had to share the Kent Coast expresses with the increasing number of Bulleid light pacifics.

Interest in train running took a dive during the war years with the absence of most train timers in war service and train services less frequent, heavier and slower. I've looked for one or two representative logs taken during the war years, most from Mr S.A.W. Harvey who was still travelling regularly on the South Eastern Section, mainly behind the various earlier 4-4-0s. I have traced a few runs on the 1.40pm Tonbridge-London Bridge, which demonstrate the conditions and general disruption of the time.

927 *Clifton*, despite the fact that wartime conditions now apply, is still in Bournemouth pristine condition with a down Waterloo-Bournemouth train at Eastleigh, 1 March 1940. J.M. Bentley Collection

1.40pm Tonbridge-London Bridge, 1941

		910 *Merchant Taylors* 304/330 tons 23.11.1941			911 *Dover* 290/310 tons 30.11.1941			916 *Whitgift* 269/290 tons 21.12.1941		
Miles	**Location**	**Times**	**Speeds**		**Times**	**Speeds**		**Times**	**Speeds**	
0.0	Tonbridge	00.00		25 L	00.00			00.00		
2.5	Hildenborough	05.41			06.03			05.59	26½	
		00.00			00.00					
4.4	Weald Interm	06.30	40		06.25	41		10.17	39	
7.4	Sevenoaks	11.38		25½ L	11.39		¾ L	16.40		¼ E
0.0		00.00			00.00			00.00		
1.5	Dunton Green	03.07			03.00			02.50		
2.8	Polehill Tnl	05.16	35		04.52	42		04.40	43	
5.5	Knockholt	09.05	42½		09.20	36		08.40	40½	
8.3	Orpington	13.22	sigs	27 L	18.15	sigs 2*		11.30	70/pws	
9.5	Petts Wood	16.15			20.35			14.40		
10.8	Chislehurst	18.40	sigs	29¼ L	22.40		7¾ L	16.45		1¾ L
13.1	Grove Park	21.50			26.00	sigs 5*		19.15	sigs 10*	
14.9	Hither Green	24.32	sigs	30½ L	29.30		10 L	22.45		3¼ L
16.5	St Johns	27.10			33.40	sigs		25.12		
17.2	New Cross	28.05		30 L	36.15	sigs	13½ L	26.49	(21 net)	4¼ L
20.2	London Bridge	32.59	(25½ net)	30½ L	45.38	sigs	18 L			

I've found one wartime run on the Western Section, the last leg of a run up from Bournemouth with one of that depot's 'Schools' which I have to say was appreciably better than my experience with 'Lord Nelsons' on a similar load commuting to Waterloo in 1958-60. So high a speed at Esher in the middle of the war was unusual and I'm not sure if such a speed was permitted at that period.

After the war, the country's railways were worn and slow to recover. The standard of track on most lines did not permit restoration to pre-war speeds for some time and in 1946 917 *Ardingly* was derailed at Catford with a Victoria-Ramsgate train with loss of life, the accident being blamed on permanent way defects. However, the 'Schools' on the Hastings line

Woking-Waterloo, 10.4.1942
7.20pm Woking (ex-Bournemouth)
924 *Haileybury* – Bournemouth
325/360 tons

Miles	**Location**	**Times**	**Speeds**	**Punctuality**
0.0	Woking	00.00		Not known
2.7	Byfleet	04.42	62	
6.3	Weybridge	07.20	59	
7.3	Walton-on-Thames	09.23	65	
8.1	Hersham	10.25	70	
9.6	Esher	-	76	
11.1	Hampton Crt Junction	12.47	75	
12.4	Surbiton	14.44		
		00.00		
2.2	New Malden	04.57	52	
4.7	Wimbledon	07.55	58	
6.4	Earlsfield	09.40	63	
8.1	Clapham Junction	11.20	40*	
9.4	Queens Road	13.30		
10.7	Vauxhall	15.57	sigs	
12.0	Waterloo	19.35	(18½ net)	Not known but schedule kept from Woking.

917 *Ardingly* after the derailment at Catford, 20 September 1946.
S.C. Townroe/Rodney Lissenden Collection

performed as well as anywhere, and in 1949 the St Leonards based 'Schools' were credited with the best performance of any Southern Region class – 2.01 minutes lost per 1,000 miles.

Nationalisation was implemented on 1 January 1948 and at that time, the 'Schools' were allocated to just four sheds:

St Leonards: 900-910
Ramsgate: 911-920
Dover: 924-927
Bricklayers Arms: 928-939

The Bricklayers Arms and St Leonards engines shared services on the Hastings route, where a number of trains, especially in the commuting hours, became very heavy and fully loaded between Tunbridge Wells and London. The performance capability of the 'Schools' was recognised in the new BR power classification 5P, the only 4-4-0 on the whole UK system to get more than a '4' rating. The 'Schools' often worked turn and turn about with Stewarts Lane and Ramsgate 'Battle of Britains' on the Folkestone, Dover and Ramsgate services and did so until the end of steam to the Kent Coast. In 1949 two 'Schools' – 30928 and 30929 – displaced the Brighton H2 'Atlantics' on the Newhaven boat trains, although they in turn lost those workings to the three SR electric locomotives, 20001-3 shortly afterwards. 30915 *Brighton*, 30918 *Hurstpierpoint* and

901 *Winchester* with multiple jet chimney and in wartime plain black livery, at Hildenborough with a Hastings-Charing Cross express, c1946.
Eric Treacy/J.M. Bentley Collection

939 *Leatherhead* at Tunbridge Wells with a Charing Cross-Hastings train, 1946. J.M. Bentley Collection

918 *Hurstpierpoint*, newly painted malachite green, enters Folkestone Junction station with a train for Dover, 25 April 1947. H.C. Casserley/ J.M. Bentley Collection

918 *Hurstpierpoint* at Sydenham Hill with a summer relief Kent Coast-Victoria train, c1947. Real Photographs/J.M. Bentley Collection

30932 *Blundells* passing through Tonbridge station and 'L' 1770 with a Hastings-Charing Cross train, 17 September 1949. Pamlin Print/J.M. Bentley Collection

30930 *Radley* were also regularly to be seen on the Newhaven boat trains later, either on reliefs or as replacements when the electric locomotives were not available. They were also the regular steeds of Royal trains in the 1950s, either to Epsom for Derby Day or to Dover for our Royals or visiting Continental dignitaries.

Throughout the 1950s, the 'Schools' continued their fine work on the Hastings route and their work on the Kent Coast lines was boosted on summer Saturdays as the British public took to fortnight holidays at the seaside. In 1951, one week in high summer (week ending 14 July) the Bricklayers Arms 'Schools' amassed 14,835 miles between seventeen engines (one of which was under repair the whole week). 30933 ran 1,515 miles and seven of them exceeded 1,000 miles in the week. In the summer timetable of 1954, at one stage there were 37 diagrams for 39 engines, plus 30915 at Stewarts Lane being standby engine for the Newhaven services.

In the summer of 1953, Ronald Nelson had a footplate run on 30932 *Blundells* with the heavy 9.15am Charing Cross-Folkestone and I give the highlights of the run together with comments made by Mr Nelson.

The timing on this train with this load was tight and despite high speeds, only on the Tonbridge-Ashford section was any appreciable amount of time regained. Ronald Nelson commented that steam pressure kept up throughout despite the hard work.

Charing Cross-Folkestone Central, summer 1953
9.15am Charing Cross
30932 *Blundells* - Bricklayers Arms
11 chs, 366/390 tons
Driver Jakes, Fireman Craddock to Ashford
Driver Rand, Fireman Boothroyd from Ashford

Miles	Location	Times	Speeds	Punctuality	Comments
0.0	Charing Cross	00.00		T	Boiler pressure 210psi
0.7	Waterloo East	02.43		¼ L	
0.0		00.00		T	Boiler pressure 220psi
		02.10/02.44 sig stand			
1.2	London Bridge	05.29		2 L	
4.2	New Cross	10.09	51	1¾ L	
4.9	St Johns	11.01	sigs 20*		
6.5	Hither Green	14.17	33	2¾ L	Full regulator 25% cut-off
10.6	Chislehurst	22.02	31/41½	4 L	
13.1	Orpington	26.02	45	5 L	22 % cut-off
14.6	Chelsfield	28.10	41½		26 % cut-off
15.9	Knockholt	30.08	36½		
19.9	Dunton Green	34.39	75		1st valve, 20% cut-off
21.4	Sevenoaks	36.32		5 L	
0.0		00.00			
3.0	Weald Box	05.41	58		
4.9	Hildenborough	07.23	75/86		1st valve, 20% cut-off
7.4	Tonbridge	09.54		5 L	
0.0		00.00			
5.3	Paddock Wood	07.37	70½	4½ L	Full ½ main, 26/20% cut-off
9.9	Marden	11.31	77/72½		
12.4	Staplehurst	13.36	76		Boiler 220psi, ½ main 24% cut-off
15.7	Headcorn	16.15	80½ /78		
20.9	Pluckley	22.45	pws 15*/49		
24.4	Chart Sdgs	26.52	66		
26.6	Ashford	29.51		3¾ L	
0.0		00.00			
4.3	Smeeth	08.03	43½		Full regulator 22 % cut-off/18 %
8.1	Westenhanger	11.41	48½		
9.3	Sandling Jcn	12.58	64/sigs 10*		
13.2	Shorncliffe	21.05	sigs	8 L	

30920 *Rugby* arrives at London Bridge with a train from Hastings, 21 April 1949. Pamlin Print/J.M. Bentley Collection

30923 *Bradfield* at Fareham with a Cardiff-Brighton train that the 'Schools' has worked from Salisbury, 25 June 1949. Pamlin Print/J.M. Bentley Collection

30929 *Malvern* at Streatham Common with a Newhaven boat train, 9 May 1949. Pamlin Print/ J.M. Bentley Collection

30929 *Malvern* at Southampton Central with a summer Saturday Bournemouth-Waterloo relief express, 2 September 1950. J.M. Bentley Collection

30920 ***Rugby*** goes tender first from Charing Cross over Hungerford Bridge back to Bricklayers Arms shed, July 1951. The photo was taken from the Festival of Britain site on the South Bank. J.M. Bentley Collection

30902 ***Wellington*** leaving Tonbridge with a Hastings-London Bridge express, April 1952. Rev A.W.V. Mace/John Scott-Morgan Collection

30929 *Malvern* at Charing Cross waiting departure for Hastings, 2 August 1952. MLS Collection

30929 ***Malvern*** prepares to stop at Folkestone Junction with a Dover-London Bridge train, c1953. Rev A.W.V. Mace/John Scott-Morgan Collection

30918 *Hurstpierpoint* at Factory Junction with a Dover-bound express, summer 1954. John Scott-Morgan Collection

30938 *St Olave's* on a down Folkestone train at Sevenoaks Tunnel, c1954. Rev A.W.V. Mace/John Scott-Morgan Collection

In 1956, Ronald Nelson made a return trip to Hastings on the footplate and his logs are shown below to give examples of the standard of work by 'Schools' in their swan song before the diesel units began to infiltrate in 1957 – at least the engine performance, although the operating on both runs left something to be desired.

Cannon Street-Hastings, September 1956
5.06pm Cannon Street Hastings
30907 *Dulwich* – St Leonards (Lemaître exhaust)
11 chs, 360/390 tons
Driver Snell, Fireman Dawson (Bricklayers Arms)

Miles	Location	Times	Speeds	Punctuality	Comments
0.0	Cannon Street	00.00		T	Heavy continuous rain
0.7	London Bridge	03.09	sigs 5*	¾ L	Boiler pressure 220psi
3.7	New Cross	07.43	53	1¼ L	½ main, 30% cut-off
6.0	Hither Green	10.16	55/51	¾ L	
10.1	Chislehurst	16.15	34/40/48½	¼ L	¾ main, 30% cut-off
14.1	Chelsfield	21.54	41		
15.4	Knockholt	23.42	37		Restrained to avoid slipping
19.4	Dunton Green	28.05	73½		
20.9	Sevenoaks	29.42	pws 15*	¾ L	
25.8	Hildenborough	36.16	74/81		
		38.25/41.09 sig stand			Up train from Dover line Xing
28.3	Tonbridge	43.26		6½ L	
31.7	High Brooms	51.27	27/24		
33.2	Tunbridge Wells	55.19		7¾ L	
0.0		00.00			
2.3	Frant	05.30	26/44		Full, 30 % cut-off
4.9	Wadhurst	08.54	53//47/55*		1st valve, 30 %
9.4	Ticehurst Road	13.44	72/83		
13.0	Etchingham	17.25		6¼ L	
0.0		00.00			
2.2	Robertsbridge	04.14	45/52		Full, 32%
4.6	MP 52	07.19	43		
5.3	Mountfield	08.11	49/54		
8.1	Battle	11.34	45		
10.1	Crowhurst	15.04		5 L	
0.0		00.00			
3.2	West St Leonards	06.12	54/sigs15*	6¼ L	
0.0		00.00			
0.9	Warrior Square	03.16		6½ L	
0.0		00.00			
0.7	Hastings	02.49		6¼ L	

The locomotive steamed well and rode superbly at speed.

Hastings-Charing Cross, September 1956
7.10pm Hastings – Charing Cross
30906 *Sherborne* – St Leonards
8chs, 255/275 tons
Driver Snell, Fireman Dawson (Bricklayers Arms)

Miles	Location	Times	Speeds	Punctuality	Comments
0.0	Hastings	00.00		1 L	Boiler pressure 220psi
0.7	Warrior Square	03.18	sigs 10*	1¼ L	
0.0		00.00			
0.9	West St Leonards	02.51		1 L	
0.0		00.00			
2.7	MP 58	06.14	34/33		
3.2	Crowhurst	07.12		2¼ L	
0.0		00.00			
2.0	Battle	04.32	44/69		Full, 30% cut-off
4.8	Mountfield	07.26	55/64		1st valve
7.9	Robertsbridge	11.06		2¼ L	
0.0		00.00			
2.2	Etchingham	03.49	49/52		
5.8	Ticehurst Road	08.19	45		Full, 30%
10.3	Wadhurst	15.57	36/ pws 25*	2¼ L	210psi
12.9	Frant	19.30	53/46/51		
15.2	Tunbridge Wells	22.51		2 L	
0.0		00.00		T	
1.5	High Brooms	03.58	pws 10*/49		
		09.11/12.01 sig stand			
4.9	Tonbridge	13.49	sigs 15*	5¾ L	220lbs psi/ Full, 32 %
7.4	Hildenborough	19.29	39/sigs 20*		
9.3	Weald s/box	24.35	28/sigs 20*/25		
12.3	Sevenoaks	31.19	pws 15*	10¼ L	
0.0		00.00			
1.5	Dunton Green	02.49	51		
2.8	Polhill Tnl	04.32	48		Full, 30% cut-off
5.5	Knockholt	08.04	46½		
6.8	Chelsfield	09.28	63		Regulator shut
8.3	Orpington	12.04	sigs 10*	11¼ L	
10.8	Chislehurst	16.22	56		
13.1	Grove Park	18.30	68/72		
14.9	Hither Green	20.07	40*	10¼ L	
17.2	New Cross	23.11	50		
20.2	London Bridge	27.35	15*	9 L	
21.4	Waterloo East	30.38		8 L	
0.0		00.00			
0.7	Charing Cross	03.28		8½ L	

Another run in pouring rain and heavily delayed by a train in front from Tonbridge, said to be a relief boat train which appeared to be doing badly (motive power unknown). 30906 was higher mileage than 30907 but rode and steamed well.

30939 *Leatherhead* departing from Ramsgate passing 'E1' 31067, c1954. Rev A.W.V. Mace/John Scott-Morgan Collection

30920 *Rugby* at Waterloo East with a down Hastings train, c1955. The first coach is one of the steel 100-seater coaches built for inclusion in electric stock sets but used in steam stock instead. Photomatic/John Scott-Morgan Collection

30901 *Winchester* winds round the curves at Frilsham, near St Leonards, with a Charing Cross-Hastings train, c1955. J.M. Bentley Collection

30908 *Westminster* heads a train for Hastings approaching West St Leonards, 6 August 1955. H. Townley/J.M. Bentley Collection

30920 ***Rugby*** at Wadhurst with a Hastings-Charing Cross train, 8 June 1955.
J.M. Bentley Collection

30920 ***Rugby*** at Bo-peep Junction leaving West St Leonards with a Tonbridge-Hastings stopping train, September 1955. A light engine ('H' 0-4-4T) stands on the line to Brighton.
J.M. Bentley Collection

30926 *Repton* departs from Ore with a train from Hastings to Ashford, 1 September 1955. Ore is the termination of the electric traction from Brighton and Hastings and the electric stock shed is seen on the right. J.M. Bentley Collection

30928 *Stowe* leaving Warrior Square for Hastings, September 1955. J.M. Bentley Collection

The first stage of the dieselisation, using diesel electric multiple unit sets after trials in the early part of 1957, was introduced in the summer timetable, June 1957. St Leonards lost all its 'Schools' allocation except 30900-30902, some went to augment the summer working to the Kent Coast and a few to Nine Elms to take over the working of the Waterloo-Lymington boat trains from the 'D15s' and short-lived mogul era (the turntable at Lymington Pier restricted the use of 4-6-0s). In June 1958, the second stage of the DEMU introduction was completed and 30900-30902 were transferred

30901 *Winchester* at West St Leonards with a train for Charing Cross, September 1956. J.M. Bentley Collection

30934 *St Lawrence* arriving at Hastings station with an express from Charing Cross, 28 August 1955. J.M. Bentley Collection

away. 'Schools' were still much in evidence on Kent Coast summer expresses and their reliefs in 1958, including Pullman car trains like the *Kentish Belle,* and on 26 July some 20 per cent of the traffic to the Kent Coast was 'Schools' hauled.

Despite the proliferation of Bulleid's pacifics and also ten BR Standard 5s (73080-9) that had arrived at Stewarts Lane in the mid-1950s, the 'Schools' could still show off their speed-worthiness on the Charing Cross-Folkestone expresses especially on the Tonbridge-Ashford section. Cecil J. Allen wrote an article 'In Praise of the Schools'

30937 *Epsom* arriving at Hastings with a train from Charing Cross, 1 September 1955. J.M. Bentley Collection

30907 *Dulwich* enters Battle station with a Charing Cross-Hastings train, c1956. J.M. Bentley Collection

in the February 1957 edition of *Trains Illustrated.* After describing a run on the 6.3pm Cannon Street-Hastings as far as Tunbridge Wells with 30909 *St Paul's,* which cleared Knockholt at 43mph, and run up to 83½ mph through Hildenborough, and 30926 *Repton* which knocked five minutes off the Tunbridge Wells-Crowhurst schedule, with a maximum of 79mph at Etchingham, he went on to publish a real 'tour de force' on the Folkestone line. 30919 *Harrow* with 395 tons gross got to Folkestone Central in 73 minutes 45 seconds on the 4.15pm *Man of Kent* from Charing Cross on 29 August 1955, with an excellent 51mph at Knockholt, 82 at Hildenborough and 77 at Staplehurst. Net time was 71¾ minutes, just over four minutes less than the demanding schedule. In the up direction in May 1957, 30935 *Sevenoaks* with 350 tons gross from a standing start at Ashford passed Tonbridge (26.6 miles) in 25 minutes 32 seconds, a gain of a minute and a half on schedule, top speeds, 80 at Headcorn and 82 at Paddock Wood. In December 1954, Driver Jakes and Lemaître 30931 had run start to stop in the down direction in 25 minutes 18 seconds with 81 at Marden and 84 after Headcorn.

Then in June 1959 came the final blow that spelt the end of the 'Schools' on main line express work in Kent. The electrification of the Chatham route to Ramsgate was completed and the new service started on 15 June 1959.

30912 *Downside* on a Ramsgate-Victoria train passing derailed 'N' 31810 at Shortlands Junction, c1956.
Ken Wightman

30932 *Blundells* at Shortlands on a Cannon Street-Hastings train diverted to start from Victoria which will take the Tonbridge line at Bickley Junction, c1956.
Ken Wightman

30931 ***King's Wimbledon*** near Walmer with the *Man of Kent* express to Margate and Ramsgate, 31 July 1956. M.R. Galley/Rodney Lissenden Collection

30917 ***Ardingly*** at Cannon Street with a train for Ramsgate and Dover, c1957. J.M. Bentley Collection

30939 ***Leatherhead*** with a Folkestone train passing through Tonbridge, c1957. MLS Collection

30903 ***Charterhouse*** with a Waterloo-Bournemouth express at Worting Junction, passing 'U' 2-6-0 31629 on Battledown flyover, December 1957. J.M. Bentley Collection

30903 *Charterhouse* at Vauxhall with a Waterloo-Lymington Pier train, 21 June 1958. Pamlin Prints/J.M. Bentley Collection

30922 *Marlborough* with a Dover-Ashford-Maidstone East train, c1958. Real Photographs/ J.M. Bentley Collection

An immediate redistribution of the 'Schools' then took place as follows:

Stewarts Lane: 30920-30923
Bricklayers Arms: 30924-30931
Ashford: 30932-30937
Dover: 30938 & 30939
Brighton: 30900, 30901, 30914, 30915
Nine Elms: 30902, 30903, 30906, 30907, 30909, 30910-30913, 30916-30919
Basingstoke: 30904, 30905, 30908.

30918 and 30923 quickly moved to Basingstoke and 30916 and 30917

30929 *Malvern* passes Factory Junction with a Victoria-Folkestone via Ashford express, 23 August 1958. R.C. Riley/ Transport Trust Collection

30920 *Rugby* departs from Cannon Street with a Hastings train, whilst a rebuilt 'West Country' waits to leave for Dover alongside a Hastings Diesel unit, c1958. R.C. Riley/Transport Trust Collection

30934 *St Lawrence* passing through Tonbridge with a Folkestone-Charing Cross train, 23 May 1959. A Hastings DEMU is in the opposite platform.
MLS Collection

30936 *Cranleigh* arriving at Chatham station with a Victoria-Ramsgate express, 22 May 1959.
MLS Collection

went back to Brighton. Then in 1960, three 'Schools' (30903, 30906 and 30909) went to Guildford for the Reading-Redhill line, and 30914, 30915 and 30916 went to Redhill itself for the same route plus a couple of remaining turns to London Bridge. The Nine Elms engines worked semi-fast trains from Waterloo to Basingstoke and Salisbury and some stopping trains west of Salisbury to Yeovil Junction.

Probably the most important express duty that remained regularly diagrammed for a 'Schools' was the pair of cross-country expresses between Birkenhead and Brighton/Margate, which the 'Schools' worked between Reading and Redhill and also the portions south of Redhill. These were substantial trains over a difficult route – especially between Guildford and Redhill, with its 1 in 96 climb to Gomshall in both directions (for gradient profile, see appendix). A couple of runs timed by A.G. Davies are shown below worked by 30924, a Bricklayers Arms engine.

Redhill-Guildford, 1961
Margate/Brighton – Birkenhead express (the 'Conti')
30924 *Haileybury* – Bricklayers Arms (Lemaître exhaust)
12 chs, 382/400 tons
28.10.1961

Miles	Location	Times	Speeds	Punctuality
0.0	Redhill	00.00		T
1.8	Reigate	04.53	33/56	
4.7	Betchworth	08.25	50	
7.3	Deepdene	11.08	61	
8.0	Dorking Town	11.54	40	T
12.8	Gomshall	19.02	30/60	
16.7	Chilworth	23.01	68	
18.5	Shalford	24.45	71	
19.2	Shalford Jcn	26.41	30*	2¼ E
20.4	Guildford	29.48		2¼ E

Guildford-Redhill, 1961
Birkenhead – Brighton/Margate express (the 'Conti')
30924 *Haileybury*
13 chs, 434/455 tons
12.8.1961

Miles	Location	Times	Speeds	Punctuality
0.0	Guildford	00.00		T
1.2	Shalford Jcn	03.24	30*	
1.9	Shalford	04.34	36	
3.8	Chilworth	07.52	28/26	
7.7	Gomshall	15.56	32/45	
12.4	Dorking Town	22.10	64	¾ E
13.2	Deepdene	22.48	60/48	
15.7	Betchworth	25.38	55	
18.6	Reigate	29.27	sig stop	
20.4	Redhill	38.46	(33½ net)	¾ L

30938 *St Olave's* at Beckenham Hill station with a relief Victoria-Ramsgate train, 9 May 1959. R.C. Riley/Transport Trust Collection

30907 ***Dulwich*** on shed at Nine Elms in the company of 'Q1' 33003 and a 'U' 2-6-0, c1959. J.M. Bentley Collection

30904 *Lancing*, allocated to Basingstoke, passes its home depot with a Salisbury-Waterloo semi-fast train, 1959. MLS Collection

Basingstoke's 30923 *Bradfield* heads an up boat train from Southampton Docks at Shawford, 7 July 1959. J.M. Bentley Collection

The final work before mass withdrawal of the remaining members in December 1962 remained the Reading-Redhill services from the Guildford and Redhill engines, London Bridge-Redhill-Brighton parcels and newspaper trains, and the Waterloo-Basingstoke semi-fasts and a few heavier commuter trains such as the 5.9pm Waterloo-Basingstoke and its equivalent up morning balancing turn, shared with Basingstoke's remaining 'King Arthurs'. The condemnation of the seventeen surviving 'Schools' in December 1962, many of them still in good condition, was a financial decision to withdraw all the remaining Maunsell designs except for the freight 'S15's by the end of the year, as there were so many Bulleid pacifics and BR Standard '4s' and '5s' still available to cover all remaining Southern Region steam passenger work.

30912 *Downside* near Winchester Junction with a Waterloo-Lymington Pier train, 16 July 1960. J.M. Bentley Collection

30918 ***Hurstpierpoint*** near Otterbourne with a Waterloo – Lymington Pier train, 16 June 1961.
J.M. Bentley Collection

30907 ***Dulwich*** departing from Basingstoke with the 12.54pm Waterloo-Salisbury semi-fast train, the regular turn of a Nine Elms Schools from 1958-1961, taken in the last year of operation, 1961.
Ken Wightman

A rundown 'Schools' near the end of its life, 30910 *Merchant Taylors,* heads a freight through Salisbury station, 25 July 1960. MLS Collection

30915 *Brighton* hauling a Brighton-Bournemouth train near Hinton Admiral, 17 July 1959. D. Robinson/MLS Collection

30936 *Cranleigh* in deplorable external condition on the 4.12pm to Margate stopping train at Paddock Wood, 30 May 1961. MLS Collection

30912 *Downside* coupled with the tender off a withdrawn 'Lord Nelson' at Stoneham with a Clapham Junction-Southampton Terminus ECS train, including two Pullman cars, 19 March 1962. J.M. Bentley Collection

30937 *Epsom* leaving Wadhurst with a Hastings train, June 1957.
Ken Wightman

30908 *Westminster* at Factory Junction with a Victoria – Dover boat train, 23 August 1958.
R.C. Riley

30926 *Repton* at Shortlands Junction heading a Ramsgate express, October 1958. Ken Wightman

30931 *King's Wimbledon* at Shortlands Junction with a Victoria-Dover via Maidstone East train, c1958. Ken Wightman

30937 *Epsom* on the *Kentish Belle* Pullman car train at Shortlands Junction, June 1958.
Ken Wightman

30937 *Epsom* at St Mary Cray Junction with a down express for the Kent Coast, 16 May 1959.
R.C. Riley

30935 *Sevenoaks* at Swans Bridge, Shortlands, with a troop train for Dover, 2 August 1959.
R.C. Riley

30937 *Epsom* at Sittingbourne with a Kent Coast express on the penultimate day of steam working on that route, 13 June 1959.
R.C. Riley

30930 *Radley* at Crowthorne with a Reading-Redhill stopping train, 25 August 1962.
Ken Wightman

30930 *Radley* at Gomshall with the Margate-Birkenhead through train (nicknamed 'the Conti') which it will work as far as Reading where it will be replaced by a Reading based 'Castle', c1962.
Ken Wightman

30925 *Cheltenham* ex-works at Eastleigh in BR mixed traffic livery, 20 March 1950. J.M. Bentley Collection

Preservation

When withdrawn, the nameplates from the condemned engines were offered to the schools concerned and many are on display in the schools' museums or other conspicuous site in the school. Three 'Schools' survived for preservation, the first, 30925 *Cheltenham,* was the official example agreed as part of the national collection under the auspices of the National Railway Museum. The prototype, 900 *Eton,* was rejected as it had been modified with the Lemaître multiple-jet exhaust, so a single chimney version was chosen instead. In 1936, a drawing was made of 925 for the RCTS as the Society was founded at Cheltenham and the Society requested this locomotive to be preserved as it had been their emblem and had been used on a number of railtours sponsored by them.

It was built in May 1934 and initially allocated to Fratton and moved to Bournemouth when the Portsmouth line was electrified in 1937. After the war, it moved to Bricklayers Arms where it resided until it was withdrawn in December 1962. It was then stored at Fratton until September 1964, when it moved to Stratford Works in East London. It was moved again in February 1968 to Preston Park, joining other former SR Maunsell engines, 30850 and 30777, selected for the national collection and awaiting restoration. It moved again to Tyseley in 1970 and once more to Dinting in 1973 where it was restored externally and called to the NRM in 1977. It was repaired and made operational for inclusion in the 'Rocket 150' celebration at Rainhill in 1980, painted in the SR malachite green livery. It was placed on display in the York museum in 1981.

It is currently at the Mid-Hants Railway having undergone overhaul by a team (led by Chris Smith) at Eastleigh Works. On completion, the locomotive featured at Railfest in June 2012 and then returned to the Mid-Hants where she is based on long term loan from the NRM. She joined fellow Maunsell Southern Railway engine 850 *Lord Nelson*.

926 *Repton* was also built in 1934 and followed the same allocations and work as 925 – Fratton and Bournemouth, then Bricklayers Arms after the Second World War.

30925 on the RCTS *Wessex-Wyvern* railtour at Beaulieu Road, 8 July 1956. L. Elsey/J.M. Bentley Collection

30925 piloting LMS 2P 4-4-0 40646 on an RCTS *East Midlander* railtour at Nottingham Midland, c1958. J.M. Bentley Collection

30925 on an RCTS 'Farewell Tour' to the 'Schools' class at Brighton, 7 October 1962. A.C. Gilbert/MLS Collection

The preserved 926 *Repton* at Steamtown Museum, Bellows Falls, Vermont, in the company of a classic GW coach and SR M7 30053, July 1967. M R Galley/R. Lissenden Collection

It was withdrawn in December 1962 and retained as enquiries of a possible purchase were made from the USA. In December 1964, it was restored to the original Southern Railway livery and in 1967 it was shipped from Liverpool to Montreal for movement on to Steamtown Museum in Vermont. It was exhibited in the open and the paintwork deteriorated and in 1974 it was repaired and placed on loan to the Cape Breton Steam Railway in Nova Scotia, Canada – equipped with cowcatcher, bell and headlight to meet safety conditions for operation of a

926 *Repton* as equipped to run on the Cape Breton Steam Railway between Glace Bay and Port Morien, Nova Scotia, 15 July 1977. M R Galley/R.Lissenden Collection

926 *Repton* at the end of its stint at the Cape Breton Steam Railway prepared for its return to Steamtown Museum, Vermont. Note the built-up tender acquired in the USA and retained on its return to the UK. M R Galley/R. Lissenden Collection

regular passenger service. In 1989 it was sold again, and returned to the UK to the NYMR, where it was again overhauled and found to be in good condition. Currently undergoing overhaul which is near completion and is planned to steam in the next few months. It is equipped with the high-sided tender acquired during its stay in North America and will appear in the SR Olive Green livery, numbered 926.

928 *Stowe* was built in 1934 at a cost of £5,000 by the Eastleigh locomotive works of the Southern Railway, and like 925 and 926, operated from Fratton, Bournemouth and Bricklayers Arms. It recorded more than a million miles of passenger service operation during 28 years of Southern main line use. It was purchased from British Railways for Lord Montagu's National Motor Museum on withdrawal in 1962. It was moved to the East Somerset Railway in 1973, and then to the Bluebell Railway where it was put into running order by the Maunsell Locomotive Society, entering service in 1981. It ran for the length of its ten-year boiler ticket and was withdrawn from service in 1991. *Stowe* was purchased by the MLS from Lord Montagu in September 2000, thus securing its future at the Bluebell. The purchase was funded in part by the sale of S15

The preserved 928 *Stowe* on the Bluebell Railway, June 1981.
Rodney Lissenden

class no. 830, which subsequently moved to the North Yorkshire Moors Railway, where it awaits restoration to working order. *Stowe* is now undergoing a full overhaul to working order, with funds being raised through the Bluebell's 'Keep Up The Pressure' campaign.

Personal Reminiscences of the 'Schools'

My first sight of a 'Schools' was a surprise. I was a schoolboy of twelve standing on Surbiton station platform waiting for my electric unit to Hampton Court and watching the usual trains that I and a gaggle of other youthful trainspotters observed each day. The up *Atlantic Coast Express* had sped by with its usual Exmouth Junction 'Merchant Navy' and the 3.54pm Waterloo had just pounded past with a Urie 'Arthur'. We awaited with more interest the Eastleigh van train which would cross over behind the 'ACE' because it was booked to be hauled by an Eastleigh engine, usually one of their many 'Scotch Arthurs', which meant the greater probability of a 'cop'. The approaching train didn't look quite right, and it was coming rather faster than usual, and suddenly we realised – it was a mythical 'Schools' (mythical because we trainspotters had never seen such a beast at Surbiton before). We couldn't believe our eyes, a malachite green 30917 *Ardingly* with large diameter multiple-jet chimney, which swept past us and was gone – so quiet compared with the usual 'King Arthur' bark. We kicked ourselves and confirmed with each other, it really was a 'Schools'. Of course, in hindsight, it was not so unusual. 'Schools' had worked through Surbiton before we were old enough to appreciate them, and would again a few years later, but in 1950 it was unusual – except for the fact that some were overhauled at Eastleigh and the Eastleigh van train was a suitable running in turn or a good way to get the engine back to the Eastern Section where it belonged.

The next 'spotting' was during a London sightseeing trip with relatives. Walking along the Embankment, I spied a 'Schools' poking out of Charing Cross canopy on Hungerford Bridge, and although I couldn't read its number which was obscured by the ironwork, I knew it was 30932 *Blundells* because I could see the unique high-sided tender.

Then around April 1953, my aunt asked me to chaperone the nine-year old daughter of a friend of hers who'd been staying in Molesey for a week during her school holidays. She had to get home to Tonbridge and I was trusted to get her from Waterloo to Waterloo East without mishap and onto a Folkestone express behind Ramsgate pacific 34078 *222 Squadron*. Having fulfilled my duty and handed Heather over to her father at Tonbridge, I just waited for the next train back, and almost immediately a Hastings-Charing Cross train ran into the platform behind the doyen of the class, lined black painted 30900 *Eton*. I regret I took no details of the journey, merely staring out of the window trainspotting, although engines behind which I'd had a run always got a distinctive identification in my Ian Allan ABC.

Most of my schooldays were spent in Godalming and the nearest railway of interest was the 'Rattler' as it was known, the Reading-Guildford section of the line to Redhill and Tonbridge. However, in the 1950s, that line had 'Ds' and SR moguls, with 'Schools' not appearing until 1959 or 60, even though one that was later stationed at nearby Guildford was 30903, named after the school where I now resided. My interest in this engine was focused on photographs I saw in *Trains Illustrated*, or one particular one I remember seeing in one of the *My Best Photographs* booklets that Ian Allan published in the late 1940s.

In November 1957, my parents moved to Woking and I commuted daily to London as I was a student at University College, Gower Street. Armed with a season ticket, I made it my task to travel behind as many steam turns to and from Waterloo as possible and in the first year my regular locomotives were Bulleid pacifics, 'Lord Nelsons', 'King Arthurs' and Standard '4' and '5' 4-6-0s. As the best source for some of the German books I had to acquire was near Charing Cross station, I sometimes caught a train across Hungerford Bridge from Waterloo East instead of walking, just to get a run behind a 'Schools'

903 *Charterhouse* at Elmstead Woods with a Charing Cross-Folkestone and Dover express, c1932. O.J. Morris/J.M. Bentley Collection

or Ramsgate 'Battle of Britain'. I can remember getting a series of 'Schools' on trains terminating from Hastings – it was just a few months before the diesel electric units took over – 30901, 30905, 30908 were among them. Once or twice I splashed out on a return ticket to London Bridge and I can remember at least once getting 30909 and returning with a Dover 'King Arthur' (30775).

On another occasion, I bought a ticket from Charing Cross to London Bridge to experience the pairing of 'L' 31760 and 30929 *Malvern*, just out of pure curiosity. The train was the 11.46am Charing Cross-Ashford via Otford, a motley collection of passenger coaches and bogie vans. In the meantime, I'd endeavoured to get a Maunsell 4-4-0 rebuild on the last Saturday of steam working on the Chatham route and instead had got a run behind 30914 *Eastbourne* on the 12.35pm Victoria-Ramsgate as far as Chatham. Unfortunately, the route was severely congested that day (as described earlier in chapters 5 and 6) and any log I might have attempted was spoilt by signal checks in all the most interesting places. I watched 30911 *Dover* bring in the 1.35pm Victoria (by now virtually on time) before seeking a return run behind an older 4-4-0 (which I got, 'L1' 31788, see last chapter).

I subsequently used the 11.46 Charing Cross several times mainly to get runs behind Ashford six-wheel tender 'King Arthurs', but on one occasion we had 30924 *Haileybury* of Bricklayers Arms, nicely decked out in Brunswick green with the latest decal on the tender. However, the Eastern Section was not on good behaviour. The stock was only platformed twenty minutes after departure time and we left with six coaches and nine vans, 348/355 tons, 26½ minutes late. We had now lost our path, of course, and suffered signal checks at Hither Green, Chislehurst, St Mary Cray and a permanent way slack to 25mph at Swanley. In between we managed 44 at Elmstead Woods, and a swift 67 between the delay at St Mary Cray and the pws at Swanley. A further 62mph before the permanent speed restriction of 25mph at Eynsford, resulted in a 35 minutes late arrival at Sevenoaks. 69mph before yet another pws (to 30mph) and station stop at Hildenborough and 51mph and another pws before Tonbridge meant we could hardly recover. We'd done well in the circumstances to be no more than 33 minutes late there.

I returned that day (4 May 1960) on the 11.30am Ramsgate-Charing Cross behind 30936 *Cranleigh* of the same depot and eight vehicles. We left 8¾ minutes late, dropped a couple of minutes to Hildenborough because of the pws to 18mph, accelerated to 36mph up to Sevenoaks Tunnel and after station overtime, left there 13¾ minutes late. We cleared Knockholt at 42mph, and after a slight signal check before Orpington, banked a steady 70-72 down to Hither Green where we caught distants on. However, a fast run in from New Cross reduced our lateness to 12 ½ minutes at Waterloo East where I alighted.

I tried the 11.46 Charing Cross a few weeks later on 1 July 1960 and got an operationally better run, although I think the loco performance of 30924 was better. 30928 *Stowe* of Bricklayers Arms had a lighter load, only three coaches plus six vans, 220 tons gross. We left Charing Cross and

30914 *Eastbourne* at Victoria with the 12.35pm to Ramsgate on the penultimate day of steam on the Chatham route to Margate and Dover, on which I travelled to Chatham, 13 June 1959. David Maidment

Waterloo East three minutes late and lost a couple more minutes to Sevenoaks, with just one signal check to 15mph at Hither Green. We recovered to 41 at Elmstead Woods, 47 at Chislehurst, 41 at Knockholt and 60 at Dunton Green. However, we were three minutes overtime at Sevenoaks and despite 66mph before Hildenborough and 52 afterwards, we were still nearly six minutes late at Tonbridge. This time the 11.30am Ramsgate had a rebuilt West Country, 34022 *Exmoor*, on its eight coach load and nearly kept time (2 late in) with 53 at Knockholt but nothing over 66mph afterwards.

However, eventually 'the mountain came to Mohammed' – after the Hastings dieselisation, completed in 1958, and towards the end of the year, the St Leonards 'Schools' migrated to the Southern's Western Section and began to appear on the Waterloo-Basingstoke/Salisbury semi-fasts, particularly the 12.54pm to Salisbury. If I had free study periods after lunch, I would make for this train, booked for a Nine Elms 'King Arthur', until one day I was surprised to find 30907 waiting for me, complete with Nine Elms shed plate. Then in quick succession 30902, 30903, 30906, 30910 and 30919 appeared on this train and I enjoyed a good number of runs behind them – virtually a dozen each behind *Wellington*, *Charterhouse* and *Dulwich*. Then after a short while, in the summer of 1959, after the Kent Coast electrification, several more appeared and Basingstoke acquired 30904, 30905, 30908, 30918 and 30923, all of which graced one of my evening trains home, the 5.09pm Waterloo-Basingstoke, a substantial crowded ten coach train. More Eastern Section 'Schools' then joined those at Nine Elms – 30911, 30912 and 30913 started appearing on the 12.54pm Waterloo. I didn't always time the runs (I did have some college work to do) but if it was becoming sprightly I'd put my papers or book away and get my watch and notebook out. However, I show below a representative group of those I timed at that stage.

Waterloo-Woking, 12.54pm Waterloo-Salisbury, 1959

	30919 *Harrow* 7 chs 29.6.1959			**30907 *Dulwich* 7 chs 31.8.1959**			**30911 *Dover* 7 chs 13.10.1959**			**30913 *Christ's Hospital* 7 chs 29.10.1959**		
Location	**Times**	**Speeds**		**Times**	**Speeds**		**Times**	**Speeds**		**Times**	**Speeds**	
Waterloo	00.00		T	00.00		T	00.00		T	00.00		T
Vauxhall	0316	35		03.08			03.08			03.16		
Queens Road	-	66		-	55		-	55		-	51	
Clapham Jcn	06.02	45*		06.28	38*		06.33	40*		06.54	40*	
Earlsfield	08.01	51		09.03	52		09.06	51		09.37	50	
Wimbledon	10.09	sigs/pws 15*		11.02	63		11.02	60		11.38	58	
New Malden	-			13.47	60		13.58	pws 15*		14.56	pws 5*	
Surbiton	16.22	73		16.08	62		18.07	46		19.00	48	
Hampton Crt Jn	-	77		-	64		19.42	60		-	62	
Esher	18.21	80		18.36	65		20.45	65		21.38	67	
Hersham	-	78		-			-	68		-	67	
Walton	20.20	76/70		21.20	60		23.17	68		24.16	66	
Weybridge	22.04	73		23.37	60		25.17	66		26.19	63	
West Weybridge	-	77		-	65		-	72		-	75	
Byfleet	24.16	74		26.11	58		27.36	69		28.39	72	
Woking	27.52 (24¾ net)	sigs sl.	3 E	29.38		1½ E	31.03 sigs sl (28 ½ net)		T	32.35 (29 net)	sigs	1½ L

Interestingly, 30919 was in the obvious worst external condition and was withdrawn early and I have no idea why I failed to time 'my' engine, 30903 *Charterhouse*, on one of the eleven runs I had behind it on this train. Perhaps I was more conscientious in completing my college work than I remember!

The Basingstoke engines worked trains that were more profitable to time, in that they were heavier commuter trains and the engine had to work harder. I show alongside a couple of runs on the morning 9.08am from Woking (due Waterloo 9.39am), followed by three on the 5.09pm evening runs.

Finally, three runs on the evening service. The usual motive power for the 9.08 up and the 5.09pm down, a balanced diagram, was a Basingstoke 'King Arthur', most often 30777, but a double-chimneyed

Woking-Waterloo, 9.08am Woking, 1959

	30904 ***Lancing***			**30918** ***Hurstpierpoint***		
	9chs, 296/330 tons			**9 chs**		
	7.10.1959			**26.10.1959**		
Location	**Times**	**Speeds**		**Times**	**Speeds**	
Woking	00.00		T	00.00		T
Byfleet	04.22	62		04.34	61	
West Weybridge	-	66		-	69	
Weybridge	07.01	60		07.02	60	
Walton	09.06	68		09.01	72	
Hersham	-	70		-	73	
Esher	11.26	73		11.20	74	
Hampton Crt Jn	12.36	64/sigs sl		12.18	72	
Surbiton	13.58	56		13.21	73	
New Malden	16.56	pws 15*		17.00	pws 5*	
Wimbledon	21.06	50		21.47	56	
Earlsfield	23.02	59		23.34	61	
Clapham Jcn	24.57	40*		25.35	sigs sl	
Queens Road	-	50		-	54	
Vauxhall	28.36	28.50				
Waterloo	31.19		¼ L	31.56		1 L
	(28½ net)			(28 net)		

30905 ***Tonbridge*** with the high-sided tender inherited from 30932 passes Woking with a Waterloo-Bournemouth summer Saturday relief train, July 1959. David Maidment

Standard '4' 4-6-0 from Basingstoke's 75076-75079 or a 'Schools' were not infrequent. 30905 *Tonbridge* with the high-sided tender was the most frequent though it was inclined to slip and was not the best performer although externally looking great (I had thirty runs behind it on the 5.09). I show on page 211 a couple of runs behind the best of the bunch, multiple-jet exhaust 30918, 30908 and an extraordinary effort behind 30923 which was on the train one night when Driver Carlisle of Basingstoke decided to stage his own locomotive trial between an 'Arthur' (30794), a Standard '4' (75078) and this 'Schools'. The 'N15' *Sir Ector* won (just), but *Bradfield*, after a slower start to Wimbledon, piled on the speed and was certainly the fastest at West Weybridge past the old Brooklands car racing track, where it was doing 83mph.

Waterloo-Woking, 5.09pm to Basingstoke, 1959 - 60

	30918 *Hurstpierpoint* 10 chs, 342/375 tons 15.3.1960			**30908 *Westminster* 10 chs 332/365 tons 3.5.1960**			**30923 *Bradfield* 10chs, 342/375 tons March 1959**		
Location	**Times**	**Speeds**		**Times**	**Speeds**		**Times**	**Speeds**	
Waterloo	00.00		T	00.00		T	00.00		T
Vauxhall	03.05			03.11			03.24		
Queen's Road	-	sigs 2*		-	50		-	55	
Clapham Jcn	08.04	40*		06.38	44*		06.46	45*	
Earlsfield	10.22	52		08.58	52		08.59	54	
Wimbledon	12.21	63		10.59	59		10.52	65	
New Malden	15.18	pws 15*		13.56	pws 20*		13.23	70	
Surbiton	19.46	47		18.08	48		15.42	72	
Hampton Crt Jn	-	62		-	59		-	74	
Esher	22.21	70		20.50	62		17.42	76	
Hersham	-	70		-	64		-	76	
Walton	24.56	69		23.28	64		19.58	77	
Weybridge	26.50	64		25.30	62		21.36	74	
West Weybridge	-	70		-	74		-	83	
Byfleet	29.11	70		27.51	70		23.33	81	
Woking	32.30		1½ L	31.02		T	26.22		4½ E
	(27¾ net)			(28 net)					
				Driver Carlisle			Driver Carlisle		

30903 *Charterhouse* leaving Woking with the 12.54pm Waterloo-Salisbury semi-fast train on which I had travelled from London, 1959. David Maidment

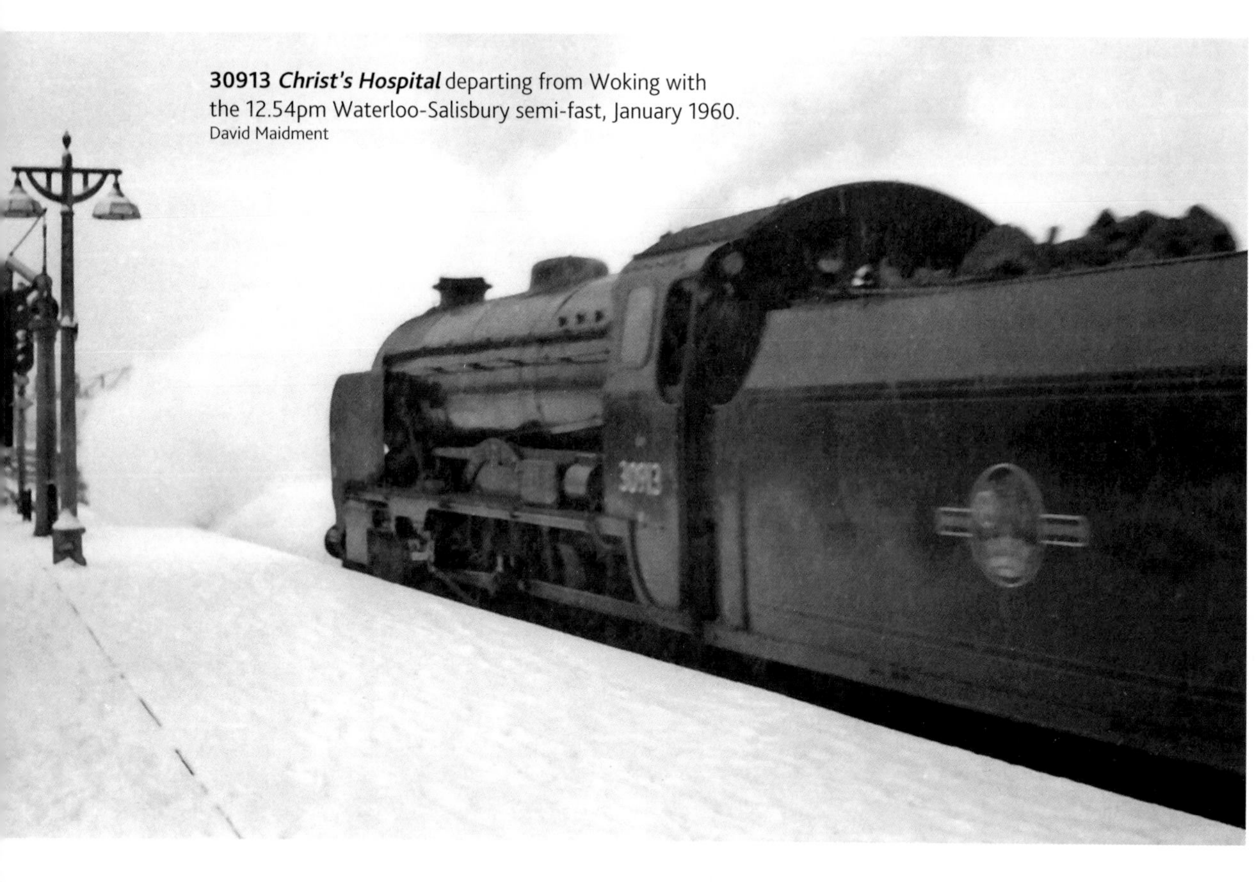

30913 *Christ's Hospital* departing from Woking with the 12.54pm Waterloo-Salisbury semi-fast, January 1960. David Maidment

Finally, I have acquired two models of the 'Schools' class – unfortunately I have found none of the Maunsell rebuild 4-4-0s, the 'Ls' or the 'L1s', a gap in the market that hopefully Hornby or Bachmann might consider filling. Here are two photos, however, of models from my own private collection, the standard Hornby model of 30934 *St Lawrence* with large diameter Lemaître multiple-jet exhaust in BR mixed traffic livery and my conversion of a Hornby 'Schools', detailed and painted by me as the school which I attended between 1951 and 1956, 30903 *Charterhouse*.

Hornby model of 30934 *St Lawrence*. David Maidment

Hornby model of 'Schools' which I customised as 30903 *Charterhouse*. David Maidment

APPENDICES

Gradient Chart – SER Route via Ashford

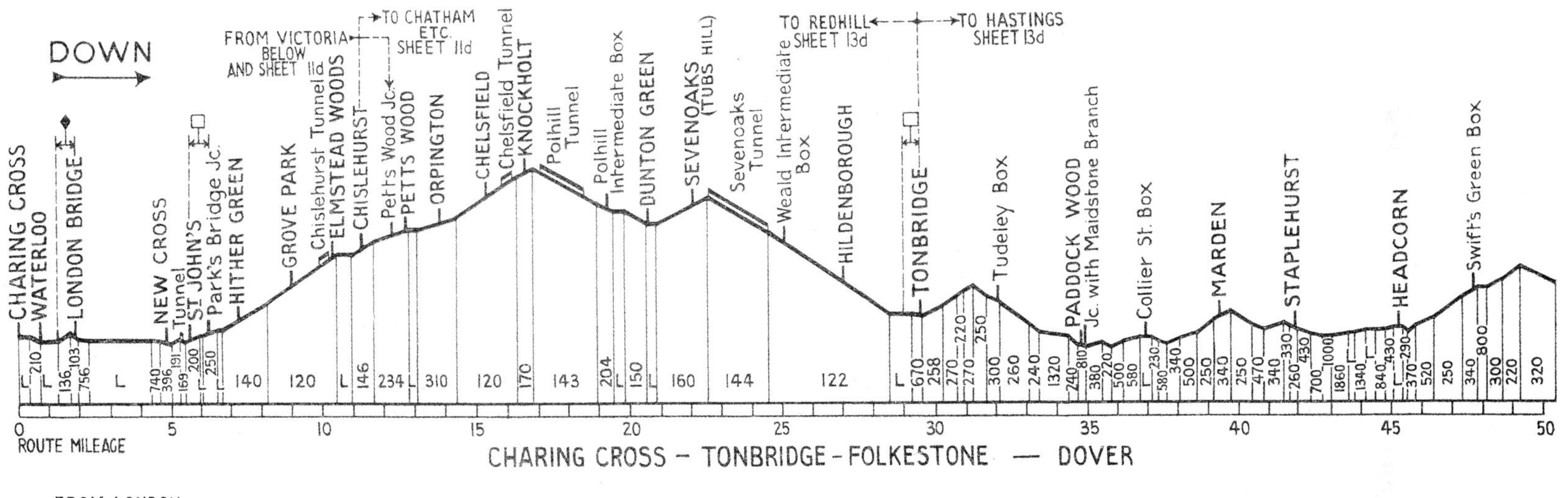

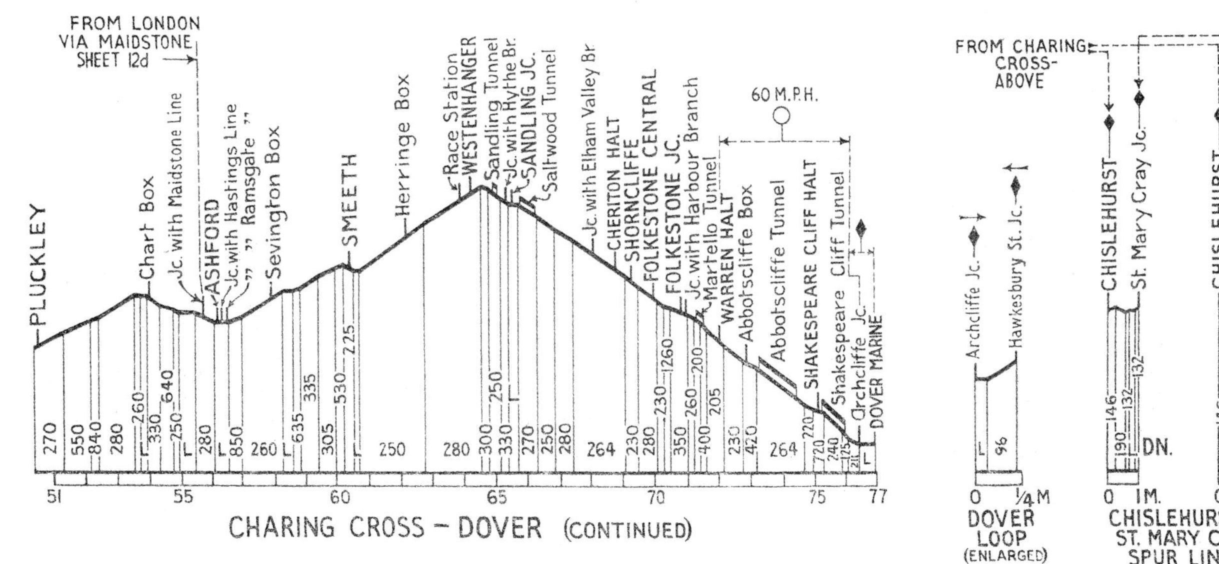

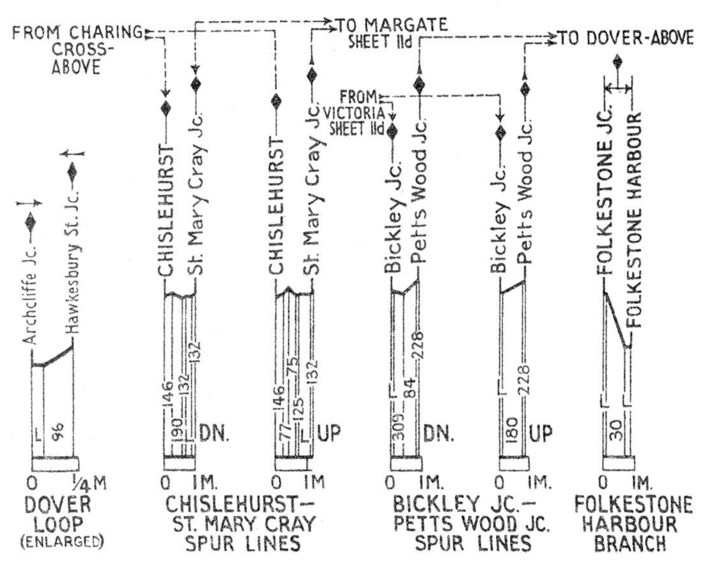

Gradient Chart – LC&DR Route via Chatham

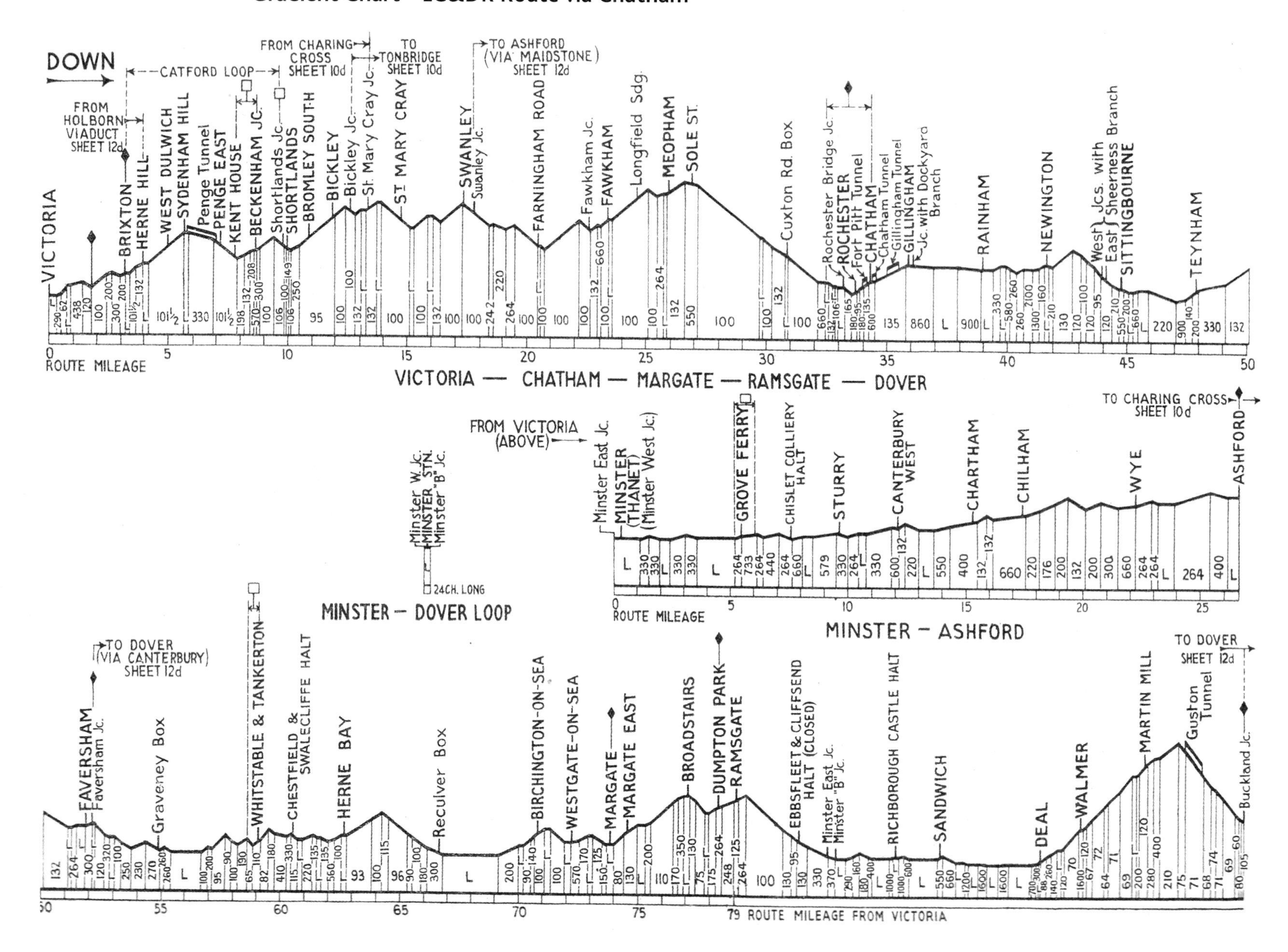

Gradient Charts – Reading-Redhill & Tonbridge-Hastings

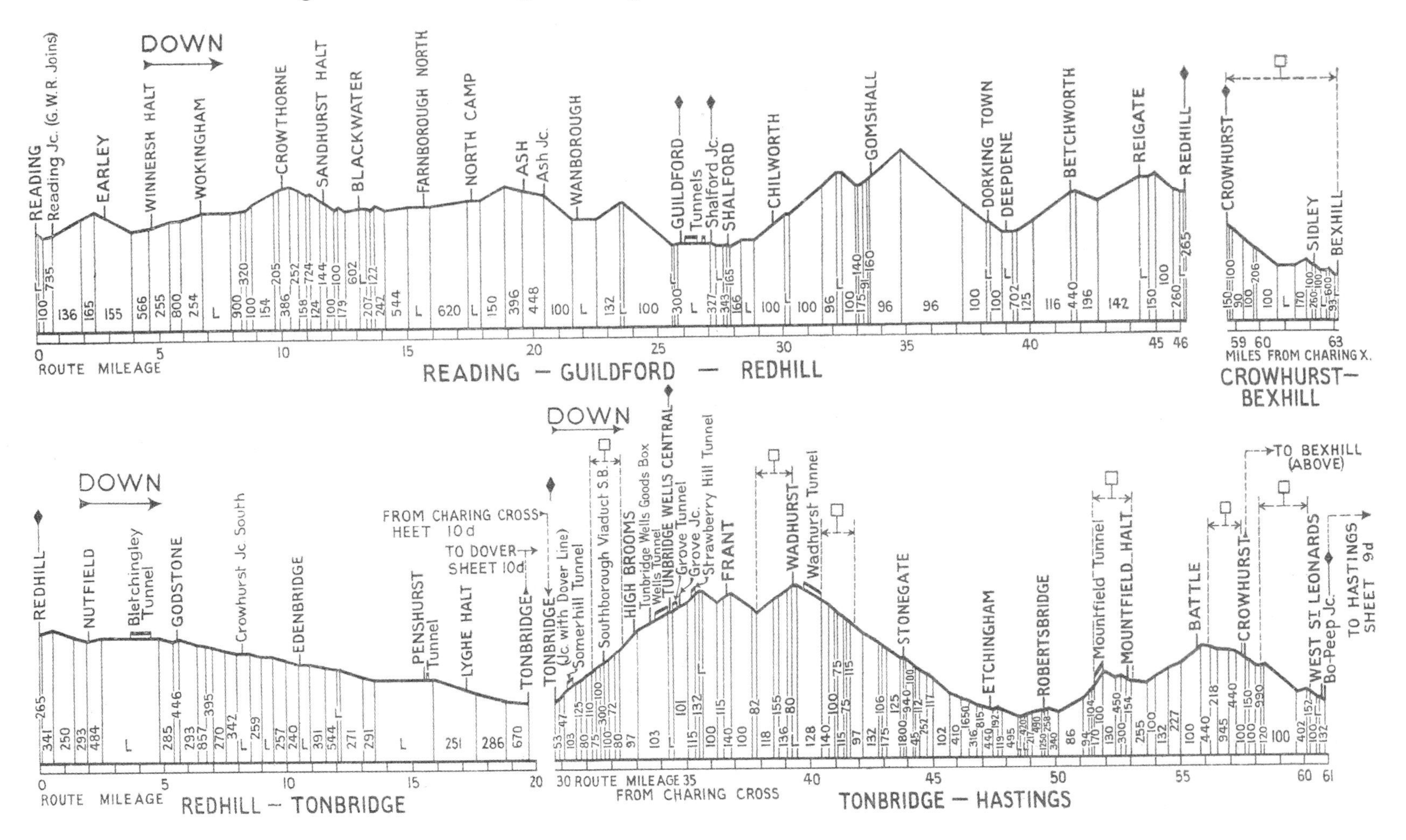

L Dimensions & Weight Diagram

Cylinders (2 inside)		20½in x 26in
Bogie wheel diameter		3ft 7in
Coupled wheels		6ft 8in
Heating surface		1,661sqft (Beyer Peacock), 1,731sqft (Borsig)
Grate area		22½sqft
Boiler pressure		160lbs psi
Weight:	Engine	57 tons 1 cwt (Beyer Peacock), 57 tons 9 cwt (Borsig)
	Tender	39 tons (Beyer Peacock), 40 tons 6 cwt (Borsig)
	Total	96 tons 1 cwt (Beyer Peacock), 97 tons 15 cwt (Borsig)
Coal capacity		4 tons
Water capacity		3,450 gallons

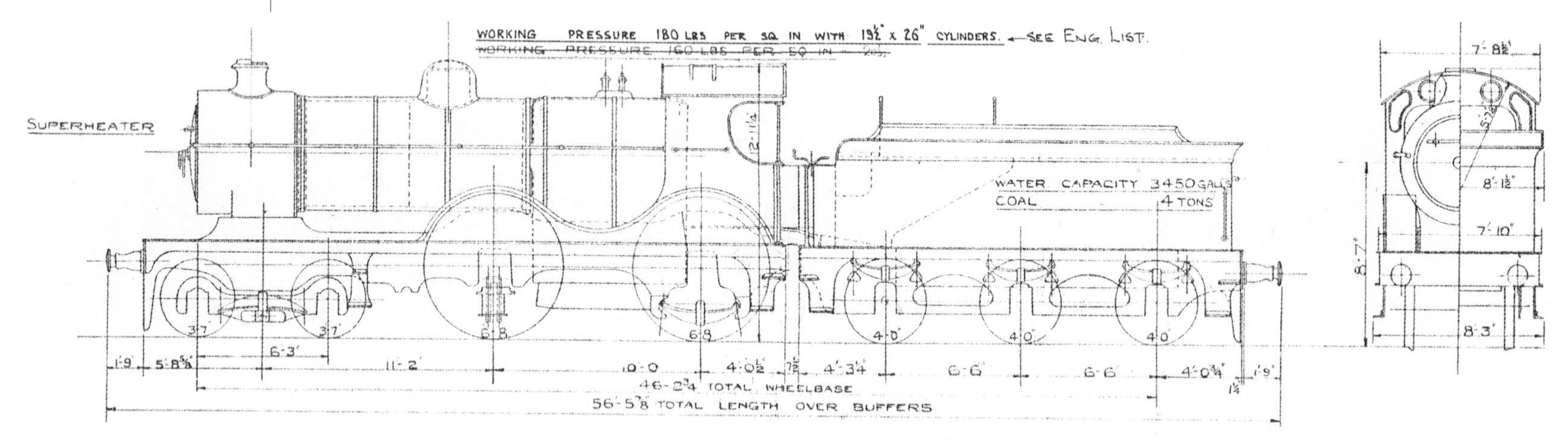

Class L 4-4-0. (Ashford dwg No.A13)

Statistics

SECR No.	Built by	SR No.	First allocation	BR No.	Last allocation	Withdrawal	Mileage
760	Beyer Peacock 8/14	1760	B. Arms	31760	Tonbridge*	6/61	
761	Beyer Peacock 8/14	1761	Dover	31761	Tonbridge	12/56	1,264,834
762	Beyer Peacock 8/14	1762	B. Arms	31762	Tonbridge*	2/60	
763	Beyer Peacock 8/14	1763	B. Arms	31763	Tonbridge*	4/60	1,402,611
764	Beyer Peacock 8/14	1764	Hastings	31764	Ramsgate*	2/61	
765	Beyer Peacock 8/14	1765	Cannon St.	31765	Faversham*	2/61	
766	Beyer Peacock 9/14	1766	Hastings	31766	Faversham*	2/61	
767	Beyer Peacock 9/14	1767	Ramsgate	31767	Faversham	10/58	
768	Beyer Peacock 9/14	1768	Ramsgate	31768	Faversham*	12/61	1,568,447
769	Beyer Peacock 9/14	1769	Hastings	31769	Ramsgate	4/56	1,119,987
770	Beyer Peacock 9/14	1770	B. Arms	31770	Tonbridge*	11/59	
771	Beyer Peacock 10/14	1771	Hastings	31771	Tonbridge*	12/61	1,244,647
772	Borsig 6/14	1772	B. Arms	31772	Tonbridge	2/59	1,461,002
773	Borsig 6/14	1773	B. Arms	31773	Tonbridge*	8/59	
774	Borsig 6/14	1774	B. Arms	31774	Tonbridge	12/58	
775	Borsig 6/14	1775	Dover	31775	Ramsgate*	8/59	

SECR No.	Built by		SR No.	First allocation	BR No.	Last allocation	Withdrawal	Mileage
776	Borsig	6/14	1776	Ashford	31776	Brighton*	2/61	
777	Borsig	6/14	1777	Dover	31777	Brighton*	9/59	
778	Borsig	6/14	1778	Dover	31778	Brighton*	8/59	
779	Borsig	7/14	1779	Dover	31779	Ramsgate*	7/59	
780	Borsig	7/14	1780	Ashford	31780	Ramsgate*	7/61	
781	Borsig	7/14	1781	Cannon St.	31781	B. Arms	6/59	1,390,562

*Allocated to Nine Elms after Kent electrification, but mostly to store.

E1 Dimensions & Weight Diagram

Cylinders (2 inside)		19in x 26in
Piston valves		10in
Valve travel (full gear)		6½in
Bogie wheels		3ft 6in
Coupled wheels		6ft 6in
Heating surfaces		1,505sqft
Grate area		24sqft
Boiler pressure		180lbs psi
Weight:	Engine	52 tons 5 cwt
	Tender	39 tons
	Total	91 tons 5 cwt
Tractive Effort (85%)		18,400lbs

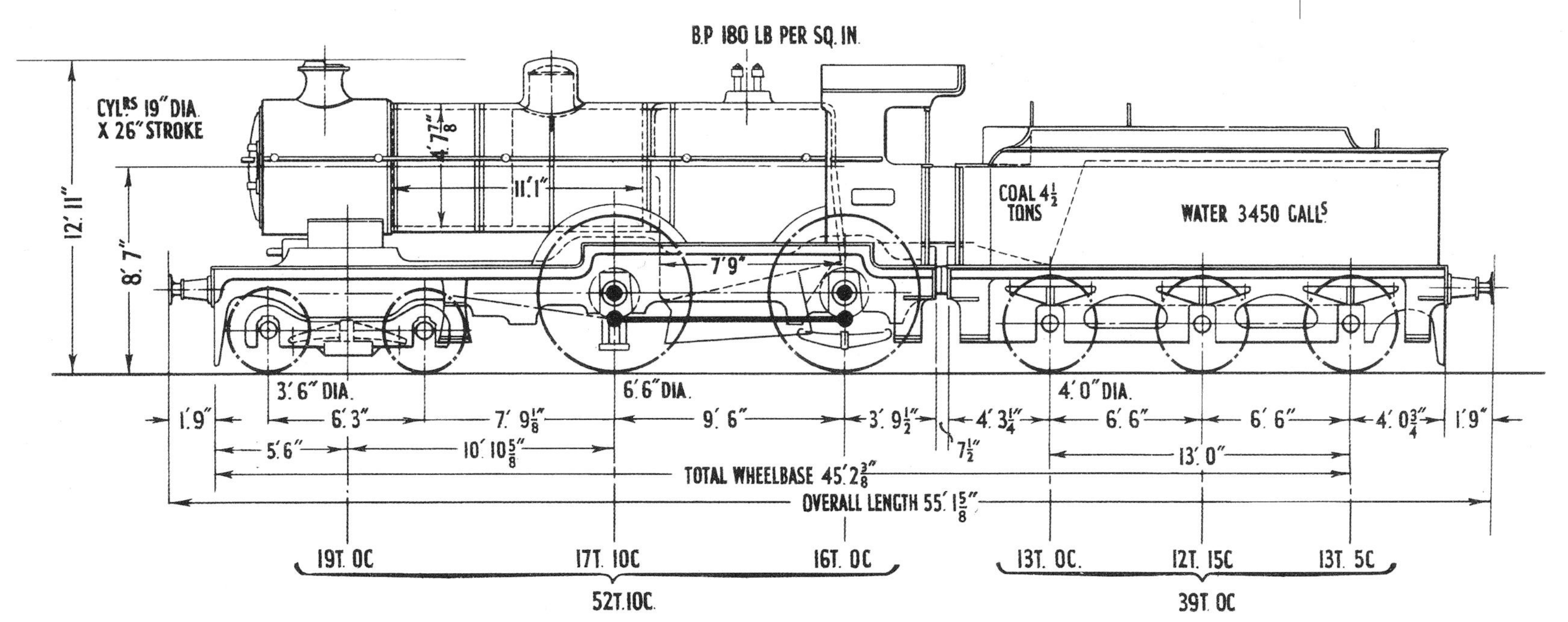

Statistics

Built as E	SR No.	Rebuilt as E1	BR No.	First allocation	Last allocation	Withdrawal	Mileage
19	1019	2/20	31019	Battersea	Stewarts Lane	4/61	
67	1067	2/20	31067	Battersea	Stewarts Lane	11/61	
160	1160	7/20	31160	Dover	B. Arms	1/51	
163	1163	4/20	31163	Dover	B. Arms	5/49	1,294,124
165	1165	5/20	31165	Battersea	B. Arms	5/59	1,571,255
179	1179	2/19	31179	Battersea	B. Arms	12/50	1,600,542
497	1497	6/20	31497	Margate West	B. Arms	10/60	
504	1504	4/20	31504	Margate West	Stewarts Lane	2/58	1,488,764
506	1506	9/20	31506	Battersea	Stewarts Lane	9/58	1,406,991
507	1507	9/20	31507	Battersea	B. Arms	7/61	
511	1511	7/20	31511	Margate West	B. Arms	12/50	

D1 Dimensions & Weight Diagram

Cylinders (2 inside)		19in x 26in
Piston valves		10in
Valve travel (full gear)		6½in
Bogie wheels		3ft 6in
Coupled wheels		6ft 8in
Heating surfaces		1,505sqft
Grate area		24sqft
Boiler pressure		180lbs psi
Weight:	Engine	51 tons 5 cwt
	Tender	39 tons
	Total	90 tons 5 cwt
Tractive Effort (85%)		18,400lbs

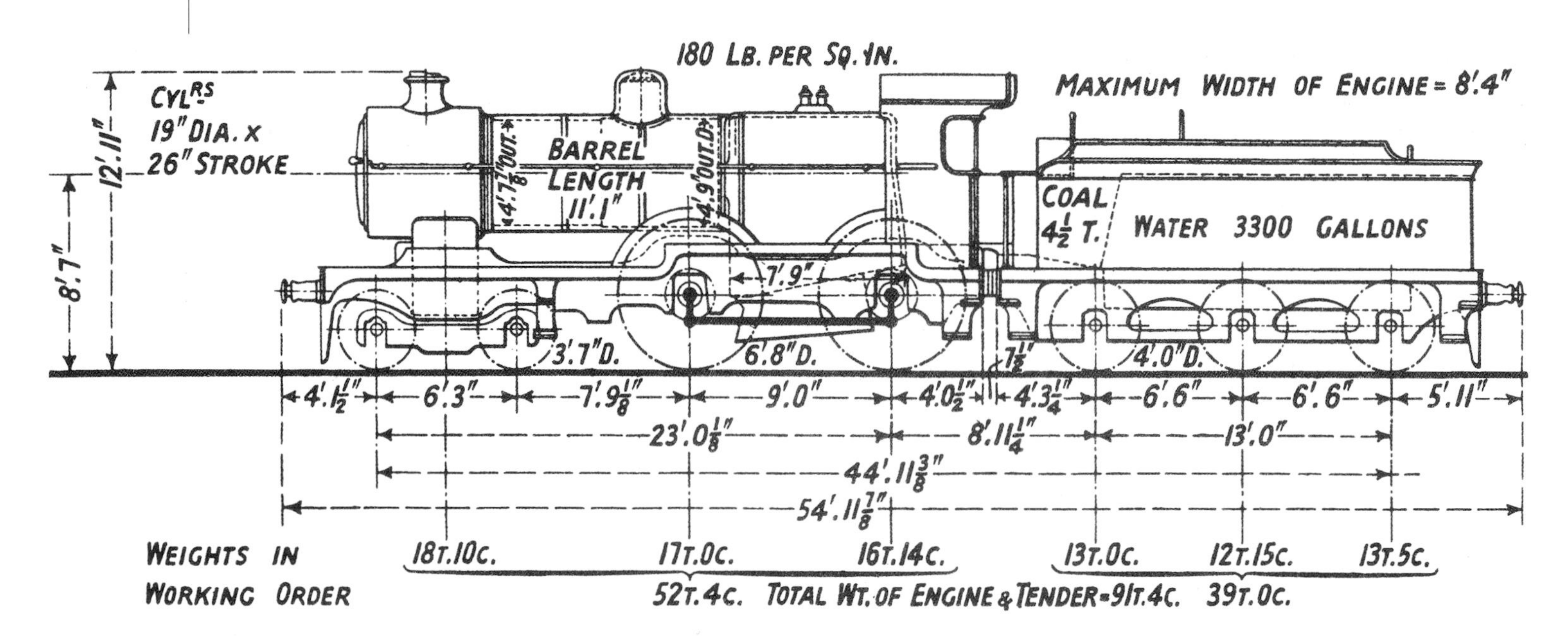

Statistics

Built as D	SR No.	Rebuilt as D1	BR No.	First allocation	Last allocation	Withdrawal	Mileage
145	1145	11/22	31145	Dover	Nine Elms	10/61	1,866,473
246	1246	4/21	31246	Margate West	Nine Elms	3/61	
247	1247	4/21	31247	Battersea	Nine Elms	7/61	1,729,426
470	1470	11/26	31470	Dover	Tonbridge	6/59	
487	1487	5/21	31487	Dover	B. Arms	2/61	
489	1489	11/21	31489	Dover	B. Arms	11/61	
492	1492	3/27	31492	Battersea	Tonbridge	1/60	
494	1494	8/21	31494	Margate West	Nine Elms	9/60	
502	1502	5/21	31502	Battersea	Faversham	2/51	
505	1505	1/27	31505	Battersea	Nine Elms	9/61	
509	1509	7/27	31509	Ramsgate	Nine Elms	5/60	
545	1545	8/21	31545	Margate West	Nine Elms	3/61	
727	1727	10/22	31727	B. Arms	Nine Elms	3/61	
735	1735	8/21	31735	Margate West	Eastleigh	4/61	1,940,163
736	1736	5/27	31736	Ramsgate	Ashford	12/50	1,306,744
739	1739	4/27	31739	Ramsgate	B. Arms	11/61	2,002,974
741	1741	2/27	31741	Ramsgate	B. Arms	9/59	1,889,511
743	1743	6/27	31743	Ramsgate	B. Arms	2/60	
745	1745	7/27	31745	Ramsgate	Stewarts Lane	2/51	
747	1747	8/21	-	Battersea	Battersea	10/44	Bomb damage
749	1749	11/21	31749	B. Arms	B. Arms	11/61	1,779,348

L1 Dimensions & Weight Diagram

Cylinders (2 inside)		19½ in x 26in
Bogie wheel diameter		3ft 7in
Coupled wheels		6ft 8in
Heating surface		1,642sqft
Grate area		22½sqft
Boiler pressure		180lbs psi
Weight:	Engine	57 tons 16 cwt
	Tender	40 tons 10 cwt
	Total	98 tons 6 cwt
Coal capacity		5 tons
Water capacity		3,500 gallons

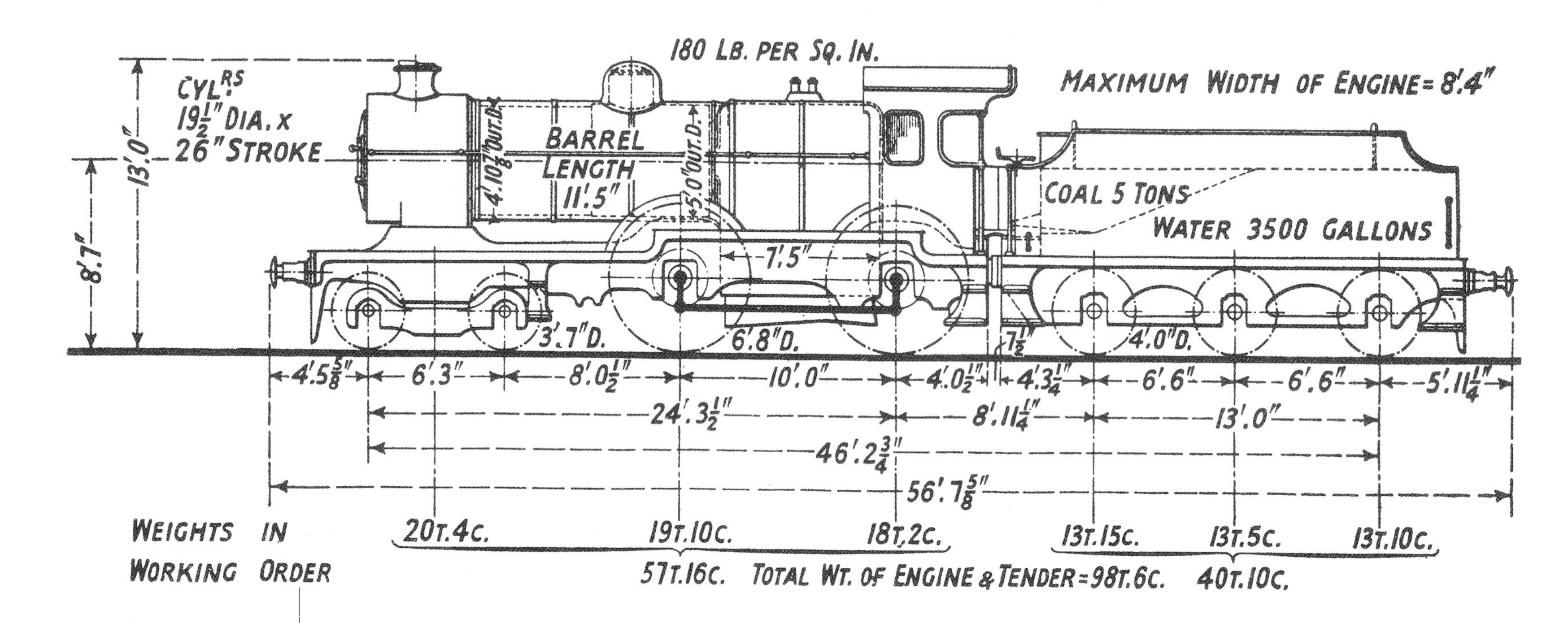

Statistics

SR No.	Built	BR No.	First allocation	Last allocation	Withdrawal	Mileage
1753	3/26	31753	Dover	Nine Elms	10/61	1,124,008
1754	3/26	31754	Dover	Nine Elms	11/61	
1755	3/26	31755	B. Arms	Ashford	8/59	1,026,608
1756	3/26	31756	Dover	Nine Elms	10/61	
1757	3/26	31757	B. Arms	Nine Elms	12/61	
1758	3/26	31758	B. Arms	Ashford	10/59	
1759	4/26	31759	B. Arms	Nine Elms	11/61	
1782	4/26	31782	Dover	Nine Elms	2/61	
1783	4/26	31783	Dover	Nine Elms	11/61	
1784	4/26	31784	Dover	Nine Elms	2/60	
1785	4/26	31785	Dover	Nine Elms	1/60	
1786	4/26	31786	B. Arms	Nine Elms	2/62	1,068,774
1787	4/26	31787	B. Arms	Nine Elms	1/61	
1788	4/26	31788	B. Arms	Nine Elms	1/60	
1789	4/26	31789	B. Arms	Nine Elms	11/61	999,423

V ('Schools')

Dimensions

Cylinders (3)	16½ in x 26in stroke
Coupled wheel diameter	6ft 7in
Bogie wheel diameter	3ft 1in
Boiler pressure	220lbs psi
Heating surface	2,049sqft

Grate area	28.3sqft
Walschaerts valve gear	
Tractive effort (85%)	25,130lbs
Axleweight	21 tons
Engine weight	67 tons 2 cwt
Tender weight	42 tons 8 cwt
Tender capacity	
coal	5 tons
water	4,000 gallons

Weight Diagram

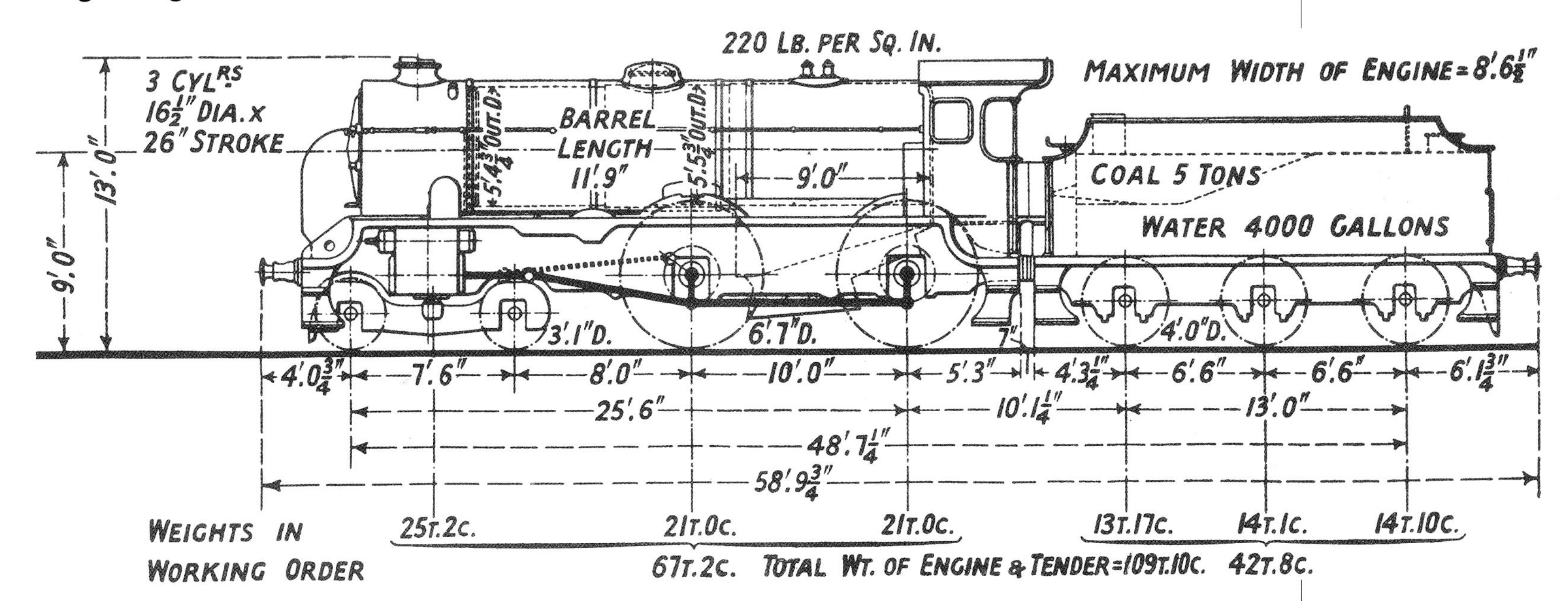

Statistics

SR No.	Built	Name	BR No.	First Allocation	Last Allocation	Withdrawal	Mileage
900	3/30	*Eton*	30900	Deal/St Leonards	Brighton	2/62	
901	3/30	*Winchester*	30901	Deal/St Leonards	Brighton	12/62	
902	4/30	*Wellington*	30902	Deal/St Leonards	Nine Elms	12/62	
903	4/30	*Charterhouse*	30903	Deal/St Leonards	Guildford	12/62	
904	5/30	*Lancing*	30904	Eastbourne/St Leonards	Basingstoke	7/61	
905	5/30	*Tonbridge*	30905	Deal/St Leonards	Basingstoke	12/61	
906	6/30	*Sherborne*	30906	Deal/St Leonards	Guildford	12/62	
907	7/30	*Dulwich*	30907	Eastbourne/St Leonards	Nine Elms	9/61	
908	7/30	*Westminster*	30908	Eastbourne/St Leonards	Basingstoke	9/61	
909	8/30	*St Paul's*	30909	Eastbourne/St Leonards	Guildford	1/62	
910	12/32	*Merchant Taylors*	30910	Eastbourne/Ramsgate	Nine Elms	11/61	
911	12/32	*Dover*	30911	Eastbourne/Ramsgate	Nine Elms	12/62	
912	12/32	*Downside*	30912	Ramsgate	Nine Elms	11/62	

SR No.	Built	Name	BR No.	First Allocation	Last Allocation	Withdrawal	Mileage
913	12/32	*Christ's Hospital*	30913	Eastbourne	Nine Elms	1/62	
914	12/32	*Eastbourne*	30914	Eastbourne	Redhill	7/61	
915	5/33	*Brighton*	30915	Eastbourne	Redhill	12/62	
916	6/33	*Whitgift*	30916	Eastbourne	Redhill	12/62	
917	6/33	*Ardingly*	30917	St Leonards	Brighton	11/62	
918	7/33	*Hurstpierpoint*	30918	St Leonards	Basingstoke	10/61	
919	7/33	*Harrow*	30919	Ramsgate	Nine Elms	1/61	
920	11/33	*Rugby*	30920	Ramsgate	Redhill	11/61	
921	11/33	*Shrewsbury*	30921	Ramsgate	Nine Elms	12/62	
922	12/33	*Marlborough*	30922	St Leonards	Stewarts Lane	11/61	
923	12/33	*Bradfield**	30923	St Leonards	Basingstoke	12/62	
924	12/33	*Haileybury*	30924	Fratton	Bricklayers Arms	1/62	
925	5/34	*Cheltenham*	30925	Fratton	Bricklayers Arms	12/62	
926	6/34	*Repton*	30926	Fratton	Bricklayers Arms	12/62	
927	6/34	*Clifton*	30927	Fratton	Bricklayers Arms	1/62	
928	6/34	*Stowe*	30928	Fratton	Bricklayers Arms	11/62	
929	8/34	*Malvern*	30929	Fratton	Bricklayers Arms	12/62	
930	12/34	*Radley*	30930	Fratton	Bricklayers Arms	12/62	
931	1/35	*King's Wimbledon*	30931	Fratton	Bricklayers Arms	9/61	
932	2/35	*Blundells*	30932	Fratton	Ashford	1/61	
933	3/35	*King's Canterbury*	30933	Fratton	Ashford	11/61	
934	3/35	*St Lawrence*	30934	Bricklayers Arms	Ashford	12/62	
935	6/35	*Sevenoaks*	30935	Bricklayers Arms	Ashford	12/62	
936	6/35	*Cranleigh*	30936	Bricklayers Arms	Ashford	12/62	
937	7/35	*Epsom*	30937	Bricklayers Arms	Ashford	12/62	
938	7/35	*St Olave's*	30938	Bricklayers Arms	Dover	7/61	965,000
939	8/35	*Leatherhead*	30939	Bricklayers Arms	Dover	6/61	

* Named *Uppingham* until 6/34

Proposed designs not built
Wainwright large bogie 4-4-0, 1907

Dimensions

Cylinders	19 ¼in x 26in
Bogie Wheels	3ft 6in
Coupled wheels	6ft 6in
Heating surface	1,577sqft
Boiler pressure	180lbs psi
Grate area	26¼sqft
Weight	
Engine	55 tons 18 cwt
Tender	39 tons 2 cwt
Water capacity	3,450 gallons
Coal capacity	4 tons

Weight Diagram

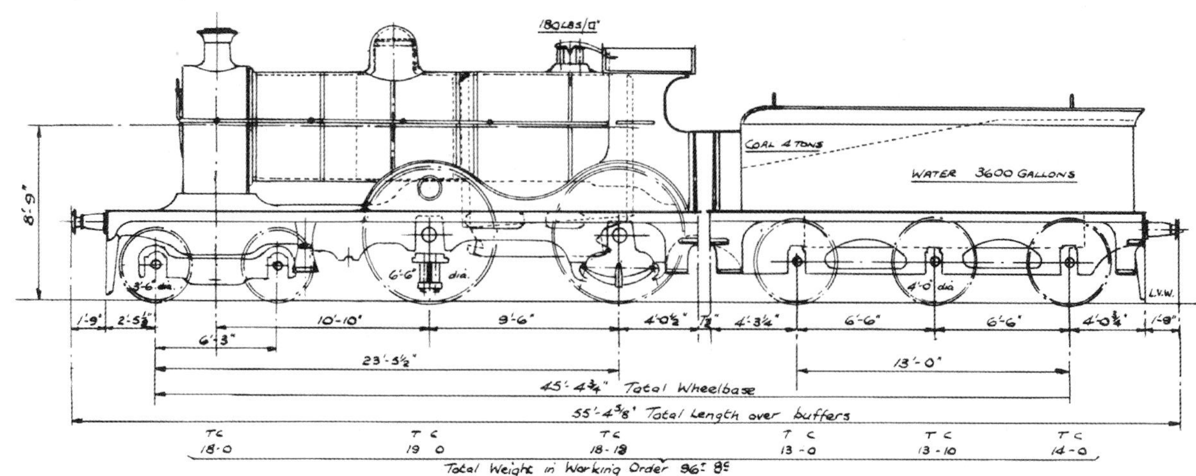

Wainwright outside cylindered six-coupled bogie 4-6-0

Dimensions

Cylinders (outside)	20¼in x 26in
Bogie wheels	3ft 6in
Coupled wheels	6ft 6in
Heating surface	2,077sqft
Grate area	26¼sqft
Boiler pressure	160lbs psi
Weight	
Engine	69 tons 0 cwt
Tender	43 tons 0 cwt
Water capacity	3,800 gallons
Coal capacity	5 tons

Weight Diagram

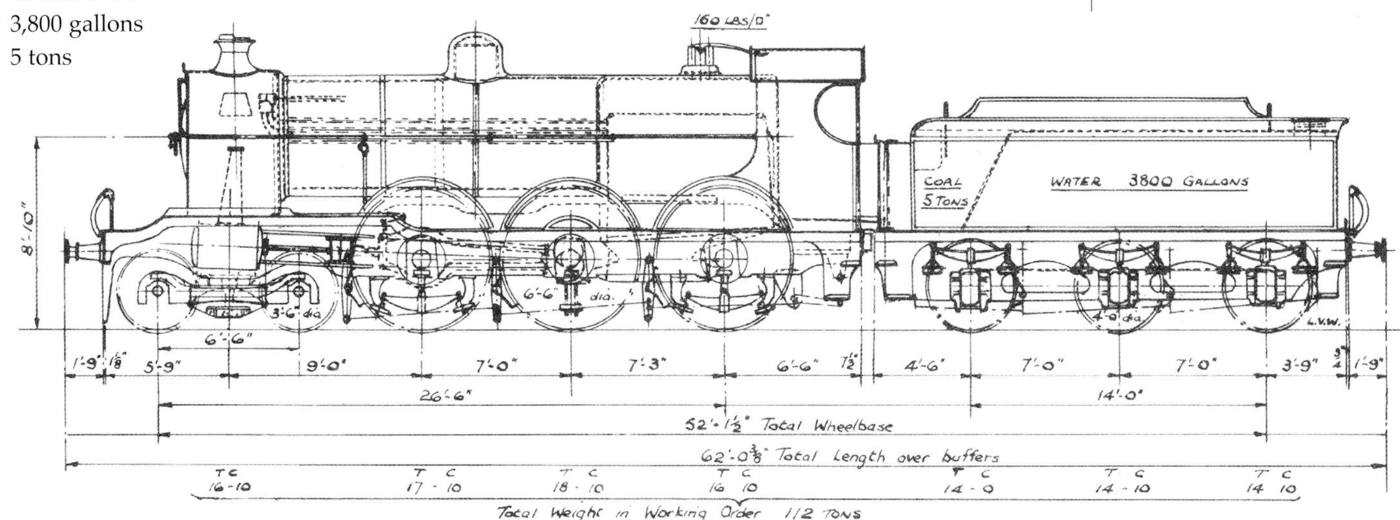

BIBLIOGRAPHY

BRADLEY, D.L., *The Locomotive History of the South Eastern & Chatham Railway,* RCTS 1980

FRYER, C.E.J., *The Rolling Rivers, Railway Monograph No.1* Platform 5, c1992

MAIDMENT, David, *Southern Urie & Maunsell 2-cylinder 4-6-0s,* Pen & Sword 2016

MAIDMENT, David, *Southern Maunsell Moguls & Tank Locomotive Classes, Pen & Sword, 2018*

NELSON, Ronald I., *Locomotive Performance, A Footplate Survey,* Ian Allan 1979

NOCK, O.S., *The Locomotives of R.E.L.Maunsell,* Edward Everard, 1954

RUSSELL, J.H., *A Pictorial Record of Southern Locomotives,* BCA/Haynes Publishing Group, 1991

SCOTT-MORGAN, John, *Maunsell Locomotives*, Ian Allan 2002

The Railway Magazine, *Gradients of the British Main-Line Railways,* Railway Publishing Company, 1947

TOWNROE, S.C., *The Arthurs, Nelsons and Schools of the Southern,* Ian Allan 1973

WINKWORTH, D.W., *The Schools 4-4-0s*, George Allen & Unwin, 1982

INDEX